BIBLIOGRAPHIC RECORD TARGET

Graduate Library
University of Michigan

Preservation Office

Storage Number: ______________________

ABV1874
UL FMT B RT a BL m T/C DT 07/18/88 R/DT 08/01/90 CC STAT mm E/L 1
010: : |a 03026353//r84
035/1: : |a (RLIN)MIUG86-B11726
035/2: : |a (CaOTULAS)160211310
040: : |a DLC/ICU |c ICU |d CStRLIN |d MiU
050/1:0 : |a QA241 |b .B78
100:1 : |a Bowden, Joseph, |d 1869-
245:00: |a Elements of the theory of integers, |c by Joseph Bowden.
260: : |a New York, |b The Macmillan company; |a London, |b Macmillan & co., ltd., |c 1903.
300/1: : |a x, 258 p. |c 20 cm.
650/1: 0: |a Number theory.
998: : |c RHJ |s 9120

ELEMENTS OF
THE THEORY OF INTEGERS

·The MM Co·

ELEMENTS OF THE

THEORY OF INTEGERS

BY

JOSEPH BOWDEN, PH.D.

PROFESSOR OF MATHEMATICS IN ADELPHI COLLEGE, BROOKLYN, NEW YORK

New York:

THE MACMILLAN COMPANY

LONDON: MACMILLAN & CO., LTD.

1903

Set up and electrotyped October, 1903

Press of
The New Era Printing Company,
Lancaster, Pa.

PREFACE.

The present volume contains the first two parts of a work which the author hopes to finish, on the Elements of the Theory of Numbers, the remaining three parts being devoted to Rational Numbers, Real Numbers, and Complex Numbers, respectivly.

The book has sprung from a desire to put the elementary theory of numbers in a logical form, starting from the three fundamental ideas of number, equality and sum, with their axioms, building up a system of theorems on these fundamental ideas, and then, by as natural a process as possible, introducing the derived ideas of greater, less, difference, integer, product, quotient, and so forth.

JOSEPH BOWDEN.

BROOKLYN, NEW YORK,
January 3, 1903.

TABLE OF CONTENTS.

PART I.

PRIMARY NUMBERS.

CHAPTER I.

FUNDAMENTAL IDEAS. AXIOMS. DEFINITIONS.
NUMBER. EQUALITY. ADDITION.

CHAPTER II.

SUBTRACTION.
GREATER. LESS. DIFFERENCE.

PART II.

INTEGERS.

CHAPTER III.

POSITIV INTEGERS. ZERO. NEGATIV INTEGERS.

CHAPTER IV.

MULTIPLICATION.

CHAPTER V.

NUMERICAL VALUE.

CHAPTER VI.

DIVISION.

DIVISIBILITY. FACTORS. QUOTIENT.

CHAPTER VII.

FACTORS.

CHAPTER VIII.

Greatest Common Factor.

CHAPTER IX.

Least Common Multiple.

CHAPTER X.

Congruence.

PART I.

PRIMARY NUMBERS.

CHAPTER I.

FUNDAMENTAL IDEAS. AXIOMS. DEFINITIONS. NUMBER. EQUALITY. ADDITION.

1. The concept of **number**, in its simplest and original sense, is a fundamental concept. It is incapable of definition, that is, it cannot be exprest in terms of ideas simpler than itself.

The concept is aroused in the mind by the consideration of groups of objects, material or mental (1), like or unlike. To every group of objects (2) belongs a number, the number of objects in the group.

2. To every single object, also, a number belongs, the number **one**.

3. Altho the word group refers properly to a collection of objects, we will, for the sake of brevity, use it to represent either a single object or many objects.

---

(1) See Locke, "An Essay Concerning Human Understanding," 1690, Bk. II., p. 84; Schubert, "Encyklopädie der Mathematischen Wissenschaften," 1898, Vol. I., p. 1.

(2) See Lefevre, "Number and its algebra," 1896, p. 19; Stieltjes, *Annales de la Faculté des Sciences de Toulouse*, 1890, Vol. IV., p. 1.

4. Notation. Each number, like any other idea, may be given a **name** (1). Thus we hav the numbers one, two, five. Numbers may also be represented by **symbols**, as 2, 7, 9.

Frequently we shall hav occasion to make a statement that is true of all numbers. In this case letters (2) of the alphabet will be used, each to represent any number we choose to consider. If a particular letter occurs more than once in a statement, it will each time stand for the same number.

5. Definition. The numbers of objects in two groups may, or may not, be the **same**. In this idea of sameness between two numbers we hav another simple, or fundamental concept. To it we giv the special name **equality**, to distinguish the sameness between numbers from the sameness between other things.

If the numbers of objects in two groups ar the same, they ar said to be **equal** (3). Representing these numbers by a and b respectivly and using the sign $=$ for the words "is equal to (4)," we may write this statement of equality $a = b$.

(1) See Ball, "A Short Account of the History of Mathematics," 1893, p. 123.

(2) Letters were sometimes used in this way to represent numbers by the ancient Greeks. The first modern European mathematician to use letters for such a purpose was Jordanus Nemorarius, "Algorithmus Demonstratus," 1534. See Schubert, "Encyklopädie," p. 5.

(3) See Tannery, "Leçons d'Arithmétique," 1900, p. 3.

(4) The sign $=$ seems to hav been used for the idea of equality first by Record, "Whetstone of Witte," 1557.

The expression $a = b$ is calld an **equality** and is read "*a* is equal to *b*" or "*a* equals *b*." It is merely an abbreviation for the statement: The number of objects in the first of the two groups under consideration is the same as the number in the second.

The number on the left of the = sign is calld the **left** or **first member** of the equality, the other number the **right** or **second member.**

If a is not equal to b, we write $a \neq b$. This expression is calld an **inequality.**

6. Axiom. *Any number is equal to itself* (1).

In symbols: $a = a$.

A longer statement of the same principle is: *The number of objects in a group is the same as the number of objects in that group.* For example, 2=2, 6=6, 9=9.

The mind grants the truth of this principle as soon as it grasps the meaning of the words in which it is stated. All such principles, with a restriction to be given later (§ 10), ar calld **axioms**(2).

7. Axiom. *If* $a = b$, *then* $b = a$(3).

In words: *If the number of objects in a group is the same as the number of objects in a second group, then the number of objects in the second group is the same as the number of objects in the first group.*

This principle also is axiomatic. It may be calld the **Commutativ Law for Equality.**

---

(1) See Stolz und Gmeiner, "Theoretische Arithmetik," 1900, p. 2.
(2) See Jevons-Hill, "Elements of Logic," 1883, p. 112.
(3) See Schubert, "Encyklopädie," p. 5; Stolz und Gmeiner, p. 2.

8. Theorem. *If* $a \neq b$, *then* $b \neq a$.

For it follows directly from § 7 that, if $b \neq a$, then $a \neq b$.

In words this says that "if a first number, which we call b, is not equal to a second number a, then the second number a is not equal to the first number b."

If we call the first number a instead of b, and call the second number b, this statement will be: If $a \neq b$, then $b \neq a$.

9. Definition. In the last two articles we hav stated two truths, each of which is as axiomatic, in the sense of § 6, as the other and each of which follows as a logical consequence from the other. It is customary in such a case not to call both of the truths axioms, but to call one an **axiom** and the other a **theorem.**

10. Definition. In general, if for any subject ther is a group of several axiomatic principles, of which one or more is a consequence of one or more of the others, it is customary to choose from this group a smaller group, such that no one of the principles in the smaller group is provable from one or more of the others in that group.

The axioms of the smaller group ar then said to be **independent** (1) of each other.

(1) See Hilbert, "The Foundations of Geometry," translated from the German by Townsend, 1902, p. 1; Moore, *Transactions of the American Mathematical Society*, 1902, Vol. III., p. 142; Huntington, *Transactions of the American Mathematical Society*, 1902, Vol. III., p. 264.

The chosen group must also be such that all the other principles in the given group and all other known principles in the subject under consideration shall be logical consequences of the axioms in the smaller group.

The axioms of the smaller group ar then said to be **sufficient** for the subject under consideration.

Usually the choice of the smaller group can be made in several ways. An endeavor should, however, be made to hav the axioms which are chosen as **simple in statement** and as **few in number** as possible.

If ther is no choice in these respects, or if these two requirements conflict with each other, the choice is made arbitrarily.

We then hav what is calld an **independent and sufficient group of axioms** for the subject under consideration.

Whether the axioms that will be set up in this chapter constitute an independent and sufficient group is left for future discussion.

That these axioms ar **consistent,** or **compatible** (1), with each other, that is, not contradictory, is evident from the fact that they ar all true of the system of numbers.

11. Definition. A **theorem** is a principle whose truth follows as a logical consequence from the axioms chosen.

---

(1) See Hilbert, *Bulletin of the American Mathematical Society*, 1902, Vol. VIII., p. 447.

Thus of the given group of principles mentiond in § 10 all those except the ones chosen for the independent and sufficient group of axioms ar theorems.

12. We will insert here a few principles of logic that will be useful in our future work.

Definition. Statements (1) ar either **unconditional** or **conditional**, the former being calld also **categorical** statements.

A categorical statement may be represented symbolically by the form "*A* is *B*."

Conditional statements ar of two kinds, **hypothetical** and **disjunctiv** (2).

A hypothetical statement may be represented by the form "If *A* is *B*, then *C* is *D*." It consists of two parts, a **hypothesis** and a **conclusion.** The hypothesis is the clause introduced by the word "if," the conclusion the clause introduced by the word "then."

A disjunctiv statement may be represented by the form "Either *A* is *B*, or *C* is *D* (3)."

13. Definition. The **contradictory** (4) of a statement is a statement to the effect that the given statement is false.

Thus the contradictory of the categorical statement "*A* is *B*" is the statement "*A* is not *B*."

---

(1) See Jevons-Hill, pp. 66, 67. (2) See Jevons-Hill, p. 150.

(3) See Jevons-Hill, p. 156.

(4) See Jevons-Hill, p. 83; Wentworth, "Plane and Solid Geometry," 1901, p. 5.

The contradictory of the hypothetical statement "If A is B, then C is D" is the statement "If A is B, then C is not D."

The contradictory of the disjunctiv statement "Either A is B, or C is D" is "Neither A is B, nor C is D," which is equivalent to "A is not B and C is not D."

14. Axiom. Evidently *a statement and its contradictory cannot both be true.*

15. Axiom. *The contradictory of the contradictory of a given statement is the original statement.*

Thus the contradictory of the contradictory of the statement "A is B" is the statement "A is not not B," or "A is B."

16. Definition. The **opposit** of a given conditional statement, hypothetical or disjunctiv, is a conditional statement of the same kind, whose parts ar respectivly the contradictories of the parts of the given statement.

Thus the opposit of the hypothetical statement "If A is B, then C is D" is the statement "If A is not B, then C is not D."

The opposit of the disjunctiv statement "Either A is B, or C is D" is the statement "Either A is not B, or C is not D."

17. Theorem. *The opposit of the opposit of a given conditional statement is the original statement.*

This follows immediately from §§ 15, 16.

18. Definition. Statements ar **simple** or **compound** (1). A **simple categorical statement** is of the form "*A* is *B*."

A **compound categorical statement** consists of two or more simple categorical statements, usually connected by the word "and." Thus "*A* is *B* and *C* is *D*" is a compound categorical statement.

19. Axiom. *The contradictory of the compound categorical statement "A is B and C is D" is the disjunctiv statement "Either A is not B, or C is not D."*

20. Definition. A **simple conditional statement** is of one of the two forms "If *A* is *B*, then *C* is *D*" and "Either *A* is *B*, or *C* is *D*."

A **compound conditional statement**, hypothetical or disjunctiv, is a conditional statement of the same kind, of which either the first or the second part is compound. Thus the statements "If *A* is *B* and *C* is *D*, then *E* is *F*," "If *A* is *B*, then *C* is *D* and *E* is *F*," and "If *A* is *B* and *C* is *D*, then *E* is *F* and *G* is *H*" ar all compound hypothetical statements; the statement "Either *A* is *B* and *C* is *D*, or *E* is *F* and *G* is *H*" is a compound disjunctiv statement.

21. Definition. The **converse** (2) of a given hypothetical statement is a hypothetical statement whose hypothesis is the conclusion of the given statement and whose conclusion is the hypothesis of the given state-

(1) See Swinton, "New English Grammar," 1881, p. 211.

(2) See Wentworth, p. 5.

ment. Thus the converse of "If *A* is *B*, then *C* is *D*" is "If *C* is *D*, then *A* is *B*."

To distinguish a given hypothetical statement from its converse, the given statement is calld the **direct** statement.

22. Axiom. *The converse of the converse of a given hypothetical statement is the original statement.*

23. Axiom. *The converse of the opposit of a given hypothetical statement is the opposit of the converse of that statement.*

24. Axiom. *If a hypothetical statement is true, the opposit of its converse is true.*

Thus, if the statement "If *A* is *B*, then *C* is *D*" is true, so also is the statement "If *C* is not *D*, then *A* is not *B*."

25. Theorem. *If the opposit of a given hypothetical statement is true, its converse is true.*

This follows directly from the preceding.

Thus, if the statement "If *A* is not *B*, then *C* is not *D*" is true, by § 24 we hav "If *C* is *D*, then *A* is *B*."

26. Theorem. *If the converse of a given hypothetical statement is true, its opposit is true.*

Thus, if the statement "If *C* is *D*, then *A* is *B*" is true, so also is the statement "If *A* is not *B*, then *C* is not *D*."

27. The theorem of § 8 follows from the axiom of § 7 by the application of the logical rule of § 24.

28. Corresponding to every hypothetical statement ther ar three others, its opposit, its converse, and the opposit of its converse. We will number these as follows:

1. Given statement.
2. Its opposit.
3. Its converse.
4. The opposit of its converse.

These four statements may be groupt in pairs, 1 and 4 together and 2 and 3 together, so that, if either one of a pair is true, the other one of that pair is also true. We may say that 1 and 4 ar equivalent to each other and 2 and 3 equivalent to each other.

29. If a hypothetical statement is true, its opposit is not necessarily true. Thus, if the statement "If *A* is *B*, then *C* is *D*" is true, so may also be the statement "If *A* is not *B*, then *C* is *D*," that is, the statement "If *A* is not *B*, then *C* is not *D*" may not be true. Similarly, if a hypothetical statement is true, its converse is not necessarily true.

30. Any hypothetical statement of the form "If *A* is *B*, then *C* is *D*" and its equivalent "If *C* is not *D*, then *A* is not *B*" ar equivalent to the disjunctiv statement "Either *A* is not *B*, or *C* is *D*."

Similarly any disjunctiv statement of the form "Either *A* is *B*, or *C* is *D*" is equivalent to either of the hypothetical statements "If *A* is not *B*, then *C* is *D*" and "If *C* is not *D*, then *A* is *B*."

31. By replacing the disjunctiv statement "Either *A* is *B*, or *C* is *D*" by its equivalent hypothetical statement, it is plain that, if this disjunctiv statement is true, the statement "Both A is B and C is D" may also be true.

If, however, the opposit disjunctiv statement "Either A is not B, or C is not D" is true, then only one of the statements "A is B" and "C is D" can be true. We can in this case say that "If A is B, then C is not D" and that "If C is D, then A is not B."

When the given disjunctiv statement and its opposit ar both true, we can say that one, and only one, of the statements "A is B" and "C is D" is true.

32. Definition. A **partial converse** of a given compound hypothetical statement is a hypothetical statement whose hypothesis contains one of the clauses of the conclusion of the given hypothetical statement and whose conclusion contains one of the clauses of the hypothesis of the given statement.

Thus a partial converse of the statement "If A is B and C is D, then E is F and G is H" is the statement "If E is F and A is B, then C is D and G is H."

33. We will now resume our discussion of the theory of numbers with the remark that the equality $a = b$ is an illustration of one of the chief causes for the great growth of the science of mathematics. This is that mathematicians have gradually developt a system of symbols by means of which the most complex mathematical ideas may be represented in a clear, compact, unambiguous form, so that the meaning is apparent at a glance.

34. Axiom. *If* $a = b$ *and* $b = c$, *then* $a = c$ (1).

In words: *If the number of objects in a group is the same as the number of objects in a second group and the latter number is the same as the number in a third group, then the number in the first group is the same as the number in the third group.*

Stated in this form the truth of this principle is self-evident.

This axiom is usually stated: *If two numbers ar equal to the same number, they ar equal to each other.*

35. Theorem. *If* $a = b$ *and* $b \neq c$, *then* $a \neq c$ (2).

For, by § 24, it follows from the preceding axiom that, if $a \neq c$, then either $a \neq b$ or $b \neq c$. § 19.

This is equivalent to the statement:

If $a = b$ and $a \neq c$, then $b \neq c$. § 30.

Or, if $b = a$ and $a \neq c$, then $b \neq c$. § 7.

From this, interchanging the letters a and b, we get the theorem.

36. Theorem. Generalization of § 34. *If we hav a series of equalities in which the right member of each equality, except the last, is also the left member of the succeeding equality, as* $a = b$, $b = c$, $c = d$, $d = e$, $\cdots$, *or, for the sake of brevity,* $a = b = c = d = e = \cdots$, (the dots $\cdots$ meaning "and so forth," that is, that ther may be any number of equalities in the series),

---

(1) See Schubert, "Encyklopädie," p. 6; Stolz und Gmeiner, p. 2.

(2) See Jevons-Hill, pp. 108, 109; Stolz und Gmeiner, p. 2.

then any number in the series is equal to any other number in the series(1).

If the two numbers to be shown equal ar next each other in the series, ther is nothing to prove; they ar given equal. If ther is one number between them, as is the case with b and d, they ar equal by the axiom, § 34.

Consider b and e, which hav two numbers between them. By the axiom $b = d$. Then, since we hav $b = d = e$, by applying the axiom again, we see that $b = e$.

So, if ther ar k numbers between the two numbers required to be proved equal, by applying the axiom once, one of the intermediate numbers disappears; by applying it k times all ar removed.

37. Definition. If we join two groups of objects so as to form a single group, the number of objects in the group thus formd is calld the **sum** (2) of the numbers of objects in the two original groups. The latter numbers are calld the **elements** of the sum.

If the elements of a sum ar a and b, we will represent the sum itself by the symbol $a + b$(3). This is read "a plus b," the sign + standing for the word "plus."

(1) See Stolz und Gmeiner, p. 2. (2) See Tannery, p. 5.

(3) The sign + was used by Widman in his "Mercantile Arithmetic," 1489. It has been conjectured that the sign was originally a warehouse mark. The first use of this sign to indicate a sum was in Stifel's "Arithmetica Integra," 1544.

38. Theorem. *If c is the sum of a and b, then $c = a + b$.*

For c and $a + b$ each represent the number of objects in the group obtaind by joining two groups of a and b objects respectivly. § 37.

Hence, since the number of objects in any group is the same as the number of objects in that group (§ 6), c is the same as $a + b$, or $c = a + b$. § 5.

39. Definition. An expression or symbol that has only one meaning is calld **determinate, univalent,** or **unambiguous.**

40. Axiom. *The sum $a + b$ of two given numbers is univalent.*

It is evident that the number of objects in the group obtaind by making a single group of two given groups of objects is determinate.

41. Axiom. $a + b = b + a$ (1).

For example, $6 + 7 = 7 + 6$.

The expressions $a + b$ and $b + a$ are two symbols representing the same idea, the number of objects in the group obtaind by joining two groups of a and b objects respectivly. The numbers a and b hav the same weight in determining the new number, their sum; they ar merely written in different order in the two expressions $a + b$ and $b + a$.

This axiom is calld the **Commutativ (2) Law for Addition.**

---

(1) This axiom seems to hav been noticed first by Gregory, "On a Difficulty in the Theory of Algebra," 1840.

(2) The word commutativ as a mathematical term was first used by Servois, Gergonne's *Annales*, 1814, Vol. V., p. 98.

42. Definition. In the idea of joining two groups of objects to form a single group, we hav a fundamental idea. This idea may fittingly be calld an **operation.** For it is the concept of an action, the joining of two groups of objects, by either a physical or a mental effort. To distinguish it from other operations we will call it the operation of **addition** (1).

The sum $a+b$ of two numbers is, then, obtaind by an operation. The sum may be calld the **result** of the operation ; the numbers a and b the **summands,** or **operands,** the things operated on ; and the sign $+$ may be regarded as a symbol to indicate what operation is performd.

The operation of addition may also be considerd from a slightly different standpoint. We may suppose that the number a is first given and that writing $+b$ after a indicates that we add b to a. In this sense a is the **operand, passiv** (2) **element,** or **augend** (3), and b the **operator, activ element,** or **increment** (4).

With this understanding the commutativ law states that *in the operation of addition the augend and the increment may be interchanged.*

43. Axiom. *If* $a=b$, *then* $a+c+b$.

This may be seen by considering three groups of a, b, and c objects respectivly, the numbers of objects in the first two being the same.

---

(1) See Schubert, "Encyklopädie," p. 6.

(2) See Schröder, "Abriss der Arithmetik und Algebra," 1874.

(3) See Schubert, "Arithmetik und Algebra," 1899, p. 17.

(4) See Schubert, "Mathematical Essays and Recreations," translated from the German by McCormack, 1898, p. 10.

This axiom might be stated: *If* $a = b$, *then* a *is not equal to any sum of which* b *is the activ element.*

44. Theorem. *If* $a = c + b$, *then* $a \neq b$.

This follows immediately from § 43 by the application of the rule of § 24.

45. Axiom. *If* $a \neq b$, *ther is some number* c, *such that either* $a = c + b$ *or* $c + a = b$.

46. Axiom. *If* $a = b$, *then* $a + c = b + c$.

This axiom might be stated: *If for the passiv element of a sum an equal number is substituted, the sum is unchanged.*

47. Axiom. *If* $a \neq b$, *then* $a + c \neq b + c$.

48. Theorem. *If* $a + c = b + c$, *then* $a = b$.

This comes immediately from § 47 by the rule of § 24.

This theorem might be stated: *If the activ elements of two equal sums ar the same, their passiv elements ar equal.*

49. Theorem. *If* $a = b$, *then* $c + a = c + b$; *and conversely, if* $c + a = c + b$, *then* $a = b$.

For the direct theorem,

we hav	$c + a = a + c.$	§ 41.
Also, since $a = b$,	$a + c = b + c.$	§ 46.
Again	$b + c = c + b.$	§ 41.
Therfor	$c + a = c + b.$	§ 36.

For the converse theorem,

we hav $a + c = c + a$ § 41.

and, by hypothesis,

$$c + a = c + b.$$

Also $c + b = b + c.$ § 41.

Hence $a + c = b + c.$ § 36.

Therfor $a = b.$ § 48.

This theorem might be stated: *If for the activ element of a sum an equal number is substituted, the sum is unchanged; and conversely, if the passiv elements of two equal sums ar the same, their activ elements ar equal.*

50. Theorem. *If $a = b$ and $c = d$, then $a + c = b + d$ (1); and conversely, if $a + c = b + d$ and $c = d$, then $a = b$; if $a + c = b + d$ and $a = b$, then $c = d$.*

For the direct theorem,

since $a = b$, $a + c = b + c.$ § 46.

Since $c = d$, $b + c = b + d.$ § 49.

Therfor $a + c = b + d.$ § 34.

This theorem is usually briefly stated: *Equals added to equals giv equals.*

For the first part of the converse theorem,
we hav, by hypothesis,

$$a + c = b + d.$$

(1) See Stolz und Gmeiner, p. 38.

2

Since $c = d$, $b + c = b + d$. § 49.

Hence $a + c = b + c$. §§ 7, 34.

Therfor $a = b$. § 48.

Similarly the second part of the converse theorem may be proved.

It will be noticed that the path taken for the proof of the converse theorem is exactly the reverse of that taken for the proof of the direct theorem. Such is frequently the case.

This theorem might also be stated: *If for the elements of a sum equal numbers ar substituted, the sum is unchanged; and conversely, if either the activ or the passiv elements of two equal sums ar equal, the other elements ar equal.*

51. Definition. Successiv Addition. After adding two numbers we may add their sum to a third, that sum to a fourth number, and so on. In order to indicate in what order the additions ar performd we will use **parentheses** ( ), [ ], or { }. The latter two kinds of parentheses ar also calld respectivly **brackets** and **braces**. Thus we may hav $(a + b) + c$, the parenthesis enclosing $a + b$ signifying that the sum $a + b$ is formd first and then c is added to that sum (1).

Similarly we may hav $((a + b) + c) + d$, $a + (b + c)$, $[a + (b + c)] + d$, $a + \{(b + c) + d\}$, $(a + b) + (c + d)$, etc.

All such expressions ar univalent.

---

(1) Parentheses were first used by Girard, 1629.

For addition is univalent. § 40.

Hence every expression obtaind by repeated addition is univalent.

Such expressions ar calld **complex sums.** The numbers a, b, c, $\cdots$, from which a complex sum is formd, ar calld its **elements.**

For the sake of distinction a sum that has only two elements is calld a **simple** sum.

52. Theorem. *If for one or more elements of a complex sum equal numbers ar substituted, the complex sum is unchanged*(1).

This follows from §§ 46, 49, 50.

For in building a complex sum we add only two numbers at a time.

53. Definition. The **standard sum** of a series of numbers in a fixt order is the sum obtaind by adding the second to the first, the third to that sum, the fourth to that, and so on.

For example, $a + b$, $(a + b) + c$, $((a + b) + c) + d$, $(((a+b)+c)+ \cdots + i) + k$ ar standard sums; wheras $(a + b) + (c + d)$ and $a + [(b + c) + (d + (e + f))]$ ar not.

54. Axiom. $(a + b) + c = a + (b + c)$(2).

This principle becomes evident if we consider three groups of a, b, c objects respectivly. In whichever of the two indicated ways the groups ar joind, the final group is the same.

(1) See Stolz und Gmeiner, p. 38.

(2) This axiom seems to hav been noticed first by Gregory.

This axiom is calld the **Associativ** (1) **Law for Addition.**

By this law, in connection with the commutativ law and §§ 46, 49, we may prove the following series of equalities: $(a+b)+c=a+(b+c)=a+(c+b)$ $=(a+c)+b=b+(a+c)=(b+a)+c=b+(c+a)$ $=(b+c)+a=c+(a+b)=(c+a)+b=c+(b+a)$ $=(c+b)+a.$

55. We now proceed to generalize the associativ law for addition in the following theorem.

Theorem. *If the order of the elements of a complex sum be kept fixt, its parentheses may be changed in any way.*

I. The sum of a standard sum and a single number is a standard sum. Thus the sum of $((a+b)+c)+d$ and e is $(((a+b)+c)+d)+e$.

II. We will now show that the sum of a single number and a standard sum is equal to the standard sum of the numbers involvd written in the same order.

Let the given sum be

$$s=a+\{([(b+c)+d]+\cdots+i)+k\}. \qquad \S\ 6.$$

Set $\quad l=([(b+c)+d]+\cdots+i)+k.$

Then $\qquad s=a+l. \qquad \S\S\ 49, 34.$

Set m for all of l except the last element k, so that

$$m=[(b+c)+d]+\cdots+i.$$

Then $\qquad l=m+k.$

And $\qquad s=a+(m+k).$

---

(1) The name "Associativ Law" was given by Wm. Rowan Hamilton, Taylor's *Philosophical Magazine*, 1844, p. 246.

By the associativ law this givs

$$s = (a + m) + k.$$

We thus hav k written in the proper place for the standard sum.

Now consider $a + m$. Setting n for all of m except the last element i, so that

$$n = [(b + c) + d] + \cdots$$

and
$$m = n + i,$$

we hav $a + m = a + (n + i) = (a + n) + i$

and $s = (a + m) + k = ((a + n) + i) + k.$

We may continue this process until the letter that stands in the inner parenthesis with a represents only a single element.

Thus

$$s = \{([(a + b) + c] + d) + \cdots + i\} + k.$$

The proof is, then, complete and s is shown to be equal to the standard sum of its elements, written in the same order.

III. We will next show that the sum of two standard sums is equal to the standard sum of all the numbers involvd, written in the same order.

Consider the sum

$$s = \{([(a + b) + c] + \cdots + e) + f\} \\ + \{([(g + h) + i] + \cdots + l) + m\}.$$

If we set

$$n = ([(a + b) + c] + \cdots + e) + f,$$

so that

$$s = n + \{([(g + h) + i] + \cdots + l) + m\},$$

we can prove in the same way as befor that

$$s = \{([(n + g) + h] + i) + \cdots + l\} + m.$$

Substituting for n what it represents, we hav

$$s = \{([(\{([(a + b) + c] + \cdots + e) + f\} + g) + h] + i) + \cdots + l\} + m.$$

IV. Next consider any complex sum whatever. This must be built up from a number of simple sums by adding these sums to single elements or to each other. Each of the simple sums we start with is a standard sum and therfor, by what has been proved above, the sum obtaind at any step of the process is equal to the standard sum of the same elements, written in the same order. Therfor the complete sum is equal to the standard sum of its elements, written in the same order.

No matter, then, how the parentheses ar arranged, the given complex sum is equal to the standard sum of the same numbers, written in the same order. Therfor all the sums obtaind by varying the arrangement of the parentheses ar the same.

56. Since a complex sum, the order of whose elements $a, b, c, d, e, \cdots$ is fixt, is independent of the arrangement of its parentheses, we may omit the parentheses without ambiguity and speak of the sum

$a + b + c + d + e + \cdots$, this expression being used to represent the sum obtaind by any arrangement of parentheses.

57. Theorem. *The sum of the left members of any number of equalities is equal to the sum of the right members.*

Let $a = b$, $c = d$, $e = f$, $g = h$, $\cdots$

The sum of the left members is $a + c + e + g + \cdots$

Substituting for $a, c, e, g, \cdots$ their respectiv equals $b, d, f, h, \cdots$, we hav

$$a + c + e + g + \cdots = b + d + f + h + \cdots \quad \S\ 52.$$

58. We will now generalize the commutativ law for addition.

Theorem. *In any complex sum the order of the elements may be changed in any way.*

Let the given sum be

$$a + \cdots + b + \cdots + c + \cdots + d + \cdots + e + \cdot$$

and the form in which we wish to write it

$$d + c + e + a + b + \cdots$$

We hav

$$a + \cdots + b + \cdots + c + \cdots + d + \cdots + e + \cdots$$

$$= ((a + \cdots + b + \cdots + c + \cdots) + d) + \cdots + e + \cdots \quad \S\ 56.$$

$$= (d + (a + \cdots + b + \cdots + c + \cdots)) + \cdots + e + \cdots \quad \S\S\ 41, 46.$$

$$= d + ((a + \cdots + b + \cdots) + c)$$
$$+ \cdots + e + \cdots \quad \S\ 55.$$
$$= d + (c + (a + \cdots + b + \cdots)) + \cdots + e + \cdots$$
$$= d + c + ((a + \cdots + b + \cdots) + e) + \cdots$$
$$= d + c + (e + (a + \cdots + b + \cdots)) + \cdots$$
$$= d + c + e + a + ((\cdots) + b) + \cdots$$
$$= d + c + e + a + (b + (\cdots)) + \cdots$$
$$= d + c + e + a + b + \cdots$$

59. The theorems of §§ 55, 58 may be combined into one as follows:

Theorem. Generalized Associativ and Commutativ Law for Addition. *In any complex sum the arrangement of the parentheses and the order of the elements may be changed in any way.*

Or, *a complex sum is independent of the arrangement of its parentheses and the order of its elements.*

60. Reverse of Addition. Separation into Parts.

To get the sum of two given numbers a and b we joind two groups of a and b objects to form a single group. Reversely, having given any group of objects, other than a group consisting of but a single object, we can divide it into two groups, which may be spoken of as **parts** of the original group, calld the **whole.** Similarly the numbers of objects in the two partial groups may be spoken of as **parts** of the number of objects in the whole group, this latter number being calld the **whole.**

61. Axiom. *Any number, except one, can be divided into two parts and is equal to the sum of those parts.*

62. Resolution into ones. Just as a number, not one, can be divided into parts, so each of its parts, not one, can be divided into parts and the process continued until each part is the number one. We hav, then, the following theorem.

63. Theorem. *Any number a, except one, is equal to the sum of a ones.*

64. Definition. On account of this property of the number one, that all other numbers are made up of ones, the number one is sometimes called **unity.** It is the **unit** or **base** upon which the system of numbers is built. For it the symbol 1 is used.

65. Definition. This property of unity may be used to form a system of symbols and names for numbers. Thus we may let 2 stand for the sum of 1 and 1, 3 for the sum of 2 and 1, and so on; thus

$$2 = 1 + 1 = 1 + 1$$
$$3 = 2 + 1 = 1 + 1 + 1$$
$$4 = 3 + 1 = 1 + 1 + 1 + 1$$
$$5 = 4 + 1$$
$$6 = 5 + 1$$
$$7 = 6 + 1$$
$$8 = 7 + 1$$
$$9 = 8 + 1 \text{ [1]}.$$

§ 38.

(1) For the origin of the symbols 1, 2, 3, 4, 5, 6, 7, 8, 9 see Ball, p. 189.

We might continue this process as far as we chose, using a distinct symbol to represent the sum of each number and unity; thus we might set

$$
\begin{aligned}
ϒ &= 9 + 1\\
∩ &= ϒ + 1\\
Ꞓ &= ∩ + 1\\
ȝ &= Ꞓ + 1\\
Ұ &= ȝ + 1\\
ɔ̄ &= Ұ + 1\\
|0 &= ɔ̄ + 1\\
&\ldots
\end{aligned}
$$

The numbers 2, 3, 4, 5, 6, 7, 8, 9, ϒ, ∩, Ꞓ, ȝ, Ұ, ɔ̄, |0 ar named two, three, four, five, six, seven, eight, nine, ten, eleven, twelv, thirteen, fourteen, fifteen, sixteen ([1]).

Since any number a is the sum of a ones, it is evident that it would hav a place somewher in such a series, if the series were carried out far enuf.

66. The system of symbols and names above given has no rule connected with it by which the numbers represented by the symbols and names may be rememberd. Thus, ther is no reason why the symbol 7, or 9, should not be used instead of the symbol 5 to represent the number 4 + 1. The meanings of the symbols and names must, then, be memorized outright.

([1]) See Pierce, "Problems of Number and Measure," 1898, p. 9.

It is evident that the further such a system is carried, the greater is the tax on the memory. Another system has therfor been devised, which avoids this difficulty. It will be explaind in a later article.

67. Definition. The system of numbers 1, 2, 3, 4, ..., each of which is equal to the preceding plus one, is calld the **natural series of numbers** (1).

68. Definition. The alternate numbers, commencing with one, in the natural series ar calld **odd**; the others **even** (2). Thus 1, 3, 5, ... ar odd; 2, 4, 6, ... even.

69. Theorem. *If a is any number in the natural series, the numbers that immediately follow a ar $a+1$, $a+2$, $a+3$, ..., $a+n$, $a+(n+1)$, ..., where 1, 2, 3, ..., n, $n+1$, ... is the natural series of numbers.*

Let the numbers that follow a in the natural series be $b, c, d, \cdots, k, l, \cdots$ Then

$b = a + 1$ § 67.

$c = b + 1 = (a + 1) + 1 = a + (1 + 1) = a + 2,$ §§ 67, 46, 54, 65, 49.

$d = c + 1 = (a + 2) + 1 = a + (2 + 1) = a + 3,$

. . . .

Moreover, if

$$k = a + n,$$

$$l = k + 1 = (a + n) + 1 = a + (n + 1).$$

Hence the numbers $b, c, d, \cdots, k, l, \cdots$ ar equal

(1) See Stolz und Gmeiner, p. 13; Euler, "Elements d'Algèbre," edition of 1807, p. 7.

(2) See Euler, p. 21.

respectivly to the numbers $a + 1$, $a + 2$, $a + 3$, $\cdots$, $a + n$, $a + (n + 1)$, $\cdots$ In this series the numbers added to a ar each, after the first, obtaind from the preceding by adding 1 to it. These numbers ar therfor the natural series of numbers.

70. Definition. Correspondence. If two groups of objects ar the same in number, we may in thot connect each object of one group with a definit object of the other group. We ar then said to make the objects of the two groups **correspond** to each other.

If one of the groups contains n objects and the other is the group of n numbers of the natural series $1, 2, 3, \cdots, n$, then, when the correspondence is made, the group of n objects is said to be **arranged in order.**

As outward evidence of the connection between the objects and the numbers, the objects may be markt with the symbols $1, 2, 3, \cdots, n$.

The object corresponding to one is calld the **first** object, the object corresponding to two, the **second,** that corresponding to three, the **third,** and so on, that corresponding to n, the **nth,** or **last** object.

Any portion of the natural series of numbers starting with 1, as $1, 2, 3, \cdots, n$, may be made to correspond to itself, each number corresponding to itself. Thus 1 is the first number in the series, 2 the second, and so on, n the last.

71. Definition. Counting (1) is the name given to

(1) The idea of number is prerequisit to, not derived from, the idea of counting. See Lefevre, p. 21.

any process for finding the number of objects in a given group.

This is usually done by separating the objects one after the other from the group and at the same time speaking, or thinking, the numbers 1, 2, 3, 4, ··· in succession until the last object has been removed. In this way we form a correspondence between the objects of the group and the natural series of numbers. The number spoken when the last object is removed is the number of objects in the group. This follows immediately from the definition of "sum" (§ 37) and the definitions of the numbers 1, 2, 3, 4, ··· (§ 65).

72. Theorem. *The number of numbers in the portion of the natural series of numbers starting with* 1 *and ending with the number a is a.*

This is evident by counting.

73. To add two given numbers a and b, we may write each number as a sum of ones and count the ones in both considerd as a single group. The result is the sum of a and b.

Or we may find the b'th number in the natural series after a. This number is $a + b$ (§ 69).

74. Addition Table (1). In order to make a table of the sums obtaind by adding all possible pairs of the numbers 1, 2, 3, ···, 10, except those sums which may not be represented by the symbols so far defined, we draw a square with its sides parallel to the edges of the paper. Then we divide the top and left-hand sides

(1) See Tannery, p. 40.

each into 10 equal parts and thru the points of division draw lines parallel to the sides forming rows and columns of little square compartments inside the given square.

The divisions of the top and left-hand sides we mark on the outside with the numbers 1, 2, 3, ···, 10, starting at the upper left-hand corner. These numbers giv the order of the rows and columns.

Now we wish to place in each compartment the number which is the sum of the two numbers which ar found respectivly at the left of the row and at the top of the column in which the compartment lies.

Thus in the first row inside the given square we wish to write the numbers that represent the sums $1+1$, $1+2$, $1+3$, ··· But these ar the same as the numbers of the natural series following 1 (§ 69). Similarly in the second row we write the numbers in the natural series after 2, and so on.

It will be noticed that this table, which is given below, is symmetrical with reference to the diagonal line starting at the upper left-hand corner and going to the lower right-hand corner. This is due to the commutativ law for addition.

This diagonal line is calld the **principal diagonal.**

Addition Table.

	1	2	3	4	5	6	7	8	9	ϒ	∩	ℭ	ƨ	Ұ	Ↄ	10
1	2	3	4	5	6	7	8	9	ϒ	∩	ℭ	ƨ	Ұ	Ↄ	10	
2	3	4	5	6	7	8	9	ϒ	∩	ℭ	ƨ	Ұ	Ↄ	10		
3	4	5	6	7	8	9	ϒ	∩	ℭ	ƨ	Ұ	Ↄ	10			
4	5	6	7	8	9	ϒ	∩	ℭ	ƨ	Ұ	Ↄ	10				
5	6	7	8	9	ϒ	∩	ℭ	ƨ	Ұ	Ↄ	10					
6	7	8	9	ϒ	∩	ℭ	ƨ	Ұ	Ↄ	10						
7	8	9	ϒ	∩	ℭ	ƨ	Ұ	Ↄ	10							
8	9	ϒ	∩	ℭ	ƨ	Ұ	Ↄ	10								
9	ϒ	∩	ℭ	ƨ	Ұ	Ↄ	10									
ϒ	∩	ℭ	ƨ	Ұ	Ↄ	10										
∩	ℭ	ƨ	Ұ	Ↄ	10											
ℭ	ƨ	Ұ	Ↄ	10												
ƨ	Ұ	Ↄ	10													
Ұ	Ↄ	10														
Ↄ	10															
10																

CHAPTER II.

SUBTRACTION.

GREATER. LESS. DIFFERENCE.

75. Definition. Having given two numbers a and b, if $a \neq b$, ther is some third number x such that either $a = x + b$ or $x + a = b$. § 45.

In the first case a is said to be **greater than** b, in the second a is said to be **less than** b (1).

More explicitly, in the first case a is said to be **x greater than** b and in the second case **x less than** b.

Thus, since $8 = 3 + 5$, 8 is greater than 5; since $3 + 5 = 8$, 5 is less than 8.

The statement "a is greater than b" is written $a > b$; the statement "a is less than b" is written $a < b$ (2).

The definitions of the terms "greater than" and "less than" may, then, be stated in symbols:

If $a = x + b$, then $a > b$; and conversely, if $a > b$, then $a = x + b$.

If $x + a = b$, then $a < b$; and conversely, if $a < b$, than $x + a = b$.

---

(1) See Tannery, p. 8; Hadamard, "Leçons de Géométrie élémentaire," p. 46.

(2) The signs $>$, $<$ seem to hav been used first by Vieta. See his "In Artem Analyticam Isagoge," 1591.

The statements $a > b$ and $a < b$ are called **inequalities.**

76. Theorem. *If* $a = x + b$ and $x \neq y$, then $a \neq y + b$.

For, since $x \neq y$, $x + b \neq y + b$. § 47.

Hence, since $a = x + b$, $a \neq y + b$. § 35.

77. Definition. From the last theorem we see that if $a = x + b$, ther is no number y, different from x, such that $a = y + b$.

Hence, if a and b ar given, x is determind by the equality $a = x + b$.

This passiv number x, which depends only on a and b and is such that $a = x + b$, we will call the **right-handed difference** between a and b.

This difference may be represented by some symbol showing the two numbers from which it is formd. The symbol $a - b$, read "a minus b," is customary (1).

The difference $a - b$ is, then, univalent, when it exists, that is, when ther is some third number x, such that $a = x + b$.

The number a may be calld the **first element** of the difference and b the **second element.**

78. Theorem. *If the right-handed difference* $a - b$ *exists and is the number* x, *then* $a - b = x$.

This follows from § 6.

79. Theorem. *If* $a = x + b$, *then* $a - b = x$; *and conversely, if* $a - b = x$, *then* $a = x + b$ (2).

(1) The sign $-$ appears first in Widman's Arithmetic. See foot-note, p. 13. It was first used in the modern sense by Stifel.

(2) See Tannery, p. 15.

80. Theorem. *If* $x + a = b$ *and* $x \neq y$, *then* $y + a \neq b$.

81. Definition. From this theorem we see that the equality $x + a = b$ determins the number x. We will call this number the **right-handed remainder** of b less a, which we will represent by the symbol $b \mid a$, which may be read "b less a."

The remainder $b \mid a$ is, then, univalent, when it exists, that is, when ther is some third number x, such that $x + a = b$.

82. Theorem. *If the right-handed remainder* $b \mid a$ *exists and is the number* x, *then* $x = b \mid a$.

83. Theorem. *If* $x + a = b$, *then* $x = b \mid a$; *and conversely, if* $x = b \mid a$, *then* $x + a = b$.

84. Definition. Implicit and Explicit Definitions. Ther ar two classes of definitions. A definition of the first class merely givs a name to an idea with which we ar already familiar, usually a fundamental idea. Such definitions ar calld **implicit** definitions. A definition of the second class expresses an idea in terms of ideas simpler than itself, either fundamental ideas or ideas that hav already been defined in terms of fundamental ideas. Such definitions ar calld **explicit** definitions.

In §§ 75, 77, 81 we hav examples of explicit definitions. The ideas "greater than," "less than," "right-handed difference," and "right-handed remainder" ar exprest in terms of "equality" and "addition."

An idea thus exprest is calld a **derived** idea.

85. We will now prove some theorems connecting the ideas represented by the symbols $>$, $<$, $-$, $|$, which hav been independently defined.

86. Theorem. *If* $a > b$, *the right-handed difference* $a - b$ *exists, and conversely.*

For, if $a > b$, $a = x + b$. § 75.

Therfor the difference $a - b$ exists and is the number x. § 77.

The converse theorem is proved by reversing the steps.

87. Hence the symbol $a - b$ as yet has no meaning, unless $a > b$.

88. Theorem. *If* $a < b$, *the right-handed remainder* $b \mid a$ *exists, and conversely.*

This is proved in the same way as § 86.

89. Theorem. *If* $a > b$, *then* $b < a$, *and* $a \mid b = a - b$; *and conversely, if* $b < a$, *then* $a > b$ (1) *and* $a - b = a \mid b$.

For, if $a > b$, $a = x + b$ and $a - b = x$. §§ 75, 77.

Hence $x + b = a$ and $x = a \mid b$. §§ 7, 81.

Therfor $b < a$ and $a \mid b = a - b$. §§ 75, 34.

The converse theorem is proved by reversing the steps.

It should be noticed that the proof of this theorem depends on the commutativ law for equality.

90. In virtue of the last theorem, whenever we hav either of the statements $a > b$, $b < a$, we may

(1) See Stolz und Gmeiner, p. 6.

replace it by the other and may replace either of the expressions $a - b$, $a \mid b$ by the other.

We will find it convenient in future never to use the symbol $\mid$, but to use in its stead $-$.

The symbols $>$ and $<$ will, however, both be useful, tho we will employ chiefly $>$. We should always bear in mind that the statement $a > b$ may at pleasure be replaced by the statement $b < a$.

91. Definition. Having given two numbers a and b, if $a \neq b$, ther is some third number x such that either $a = b + x$ or $a + x = b$. §§ 45, 41.

In the first case a may be said to be **larger than** b, in the second case **smaller than** b.

The statement "a is larger than b" may be written $a \succ b$; the statement "a is smaller than b" $a \prec b$.

The definitions of the terms "larger than" and "smaller than" may, then, be stated in symbols:

If $a = b + x$, then $a \succ b$; and conversely, if $a \succ b$, then $a = b + x$.

If $a + x = b$, then $a \prec b$; and conversely, if $a \prec b$, then $a + x = b$.

92. Theorem. *If $a = b + x$ and $x \neq y$, then $a \neq b + y$.*

93. Definition. From this theorem we see that the equality $a = b + x$ determins the number x.

This activ number x, which depends only on b and a and is such that $a = b + x$, we will call the **left-handed difference**(1) between b and a. For this differ-

(1) See Grassmann, "Ausdehnungslehre," edition of 1878, p. 5; Stolz und Gmeiner, p. 5; Schubert, "Encyklopädie," p. 9.

ence we will use the symbol $b \underset{\cdot}{-} a$, which may be read "b from a."

The difference $b \underset{\cdot}{-} a$ is, then, univalent, when it exists, that is, when ther is a third number x, such that $a = b + x$.

94. Theorem. *If the left-handed difference $b \underset{\cdot}{-} a$ exists and is the number x, then $b \underset{\cdot}{-} a = x$.*

95. Theorem. *If $a = b + x$, then $b \underset{\cdot}{-} a = x$; and conversely, if $b \underset{\cdot}{-} a = x$, then $a = b + x$.*

96. Theorem. *If $a + x = b$ and $x \neq y$, then $a + y \neq b$.*

97. Definition. From this theorem we see that the equality $a + x = b$ determins the number x. We will call this number the **left-handed remainder** when a is taken from b. For this remainder we will use the symbol $a \mid\!\cdot\, b$, which may be read "a sul b."

The remainder $a \mid\!\cdot\, b$ is, then, univalent when it exists, that is, when ther is a third number x, such that $a + x = b$.

98. Theorem. *If the left-handed remainder $a \mid\!\cdot\, b$ exists and is the number x, then $x = a \mid\!\cdot\, b$.*

99. Theorem. *If $a + x = b$, then $x = a \mid\!\cdot\, b$; and conversely, if $x = a \mid\!\cdot\, b$, then $a + x = b$.*

100. Theorem. *If $a \succ b$, the left-handed difference $b \underset{\cdot}{-} a$ exists, and conversely.*

101. Theorem. *If $a \prec b$, the left-handed remainder $a \mid\!\cdot\, b$ exists, and conversely.*

102. Theorem. *If $a > b$, then $a \succ b$ and $b \underset{\cdot}{-} a = a - b$; and conversely, if $a \succ b$, then $a > b$ and $a - b = b \underset{\cdot}{-} a$.*

The proof of this theorem depends on the commutativ law for addition.

103. Theorem. *If* $a < b$, *then* $a \prec b$ *and* $a \mid\!\cdot\, b = b - a$; *and conversely, if* $a \prec b$, *then* $a < b$ *and* $b - a = a \mid\!\cdot\, b$.

The proof of this theorem also depends on the commutativ law for addition.

104. Definition. In virtue of the last two theorems we will in future never use the symbols $\succ$ and $\prec$, but will use in their stead $>$ and $<$ respectivly. Similarly instead of $b \mathbin{\underset{\cdot}{-}} a$ and $b \mid\!\cdot\, a$ we will use $a - b$.

In virtue of §§ 89, 102, 103 we will in future instead of the four terms "right-handed difference," "right-handed remainder," "left-handed difference," and "left-handed remainder" use the single term "**difference.**"

105. Theorem. *No number is greater than itself.*

For, since $a = a$, $a \mp x + a$. § 43.

106. Definition. The symbols $\ngtr$ and $\nless$ ar used to mean "is not greater than" and "is not less than" respectivly.

The expression $a \geqq b$ means that either $a = b$ or $a > b$. Similarly the expressions $a \leqq b$, $a \underset{<}{\overset{>}{=}} b$, etc., ar defined.

107. Definition. Complex Expressions containing Plus and Minus Signs. The difference of two numbers may be connected with other numbers by plus or minus signs, provided always that the first element of every difference is greater than the second element.

Thus we may hav expressions like $[(a-b)+c]-(d-e)$ or $\{[(a+b)-c]+d\}-(e+f)$.

All such expressions ar univalent, since every sum and every difference is univalent. §§ 40, 77.

108. Theorem. $a+b>b$ (1) and $(a+b)-b=a$ (2).

For $a+b=a+b$.

Therfor $a+b>b$ and $(a+b)-b=a$. §§ 75, 79.

109. Theorem. *A whole is greater than each of its parts.*

110. Theorem. *If $a>b$, then $(a-b)+b=a$.*

Since $a>b$, $a=x+b$ and $a-b=x$. §§ 75, 79.

Hence $(a-b)+b=x+b$. § 46.

Therfor $(a-b)+b=a$. § 34.

111. Theorem. *If $a>b$, then $a+c>b+c$ and $(a+c)-(b+c)=a-b$; and conversely, if $a+c>b+c$, then $a>b$ and $a-b=(a+c)-(b+c)$.*

For the direct theorem,

since $a>b$, $a=x+b$ and $a-b=x$. §§ 75, 79.

Hence $a+c=(x+b)+c=x+(b+c)$ §§ 46, 54.

and $(a+c)-(b+c)=x$. §§ 34, 79.

Therfor $a+c>b+c$ § 75.

and $(a+c)-(b+c)=a-b$. § 34.

---

(1) See Stolz und Gmeiner, p. 15.

(2) See Schubert, "Encyklopädie," p. 9; Tannery, p. 44.

For the converse theorem,

since $a + c > b + c$, $a + c = x + (b + c) = (x + b) + c$
and $(a + c) - (b + c) = x$.

Hence $a = x + b$ and $a - b = x$.

Therfor $a > b$ and $a - b = (a + c) - (b + c)$.

112. Theorem. *If $a > b$, then $c + a > c + b$ and $(c + a) - (c + b) = a - b$; and conversely, if $c + a > c + b$, then $a > b$ and $a - b = (c + a) - (c + b)$.*

113. Theorem. *If $a = b$ and $b > c$, then $a > c$ and $a - c = b - c$; and conversely, if $a > c$, $b > c$, and $a - c = b - c$, then $a = b$.*

For the direct theorem,

Since $b > c$, $b = x + c$ and $b - c = x$. §§ 75, 79.

Hence, since $a = b$, $a = x + c$ and $a - c = x$.

Therfor $a > c$ and $a - c = b - c$.

For the converse theorem, let $a - c = x$.
Then $b - c = x$.
Hence $a = x + c$

and $b = x + c$.
Therfor $a = b$.

114. Theorem. *If $a > b$ and $b = c$, then $a > c$ and $a - b = a - c$; and conversely, if $a > b$, $a > c$, and $a - b = a - c$, then $b = c$.*

For the direct theorem,

since $a > b$, $a = x + b$ and $a - b = x$.

Since $b = c$, $x + b = x + c$.

Hence $a = x + c$ and $a - c = x$.

Therfor $a > c$ and $a - b = a - c$.

For the converse theorem, let $a - b = x$. Then $a - c = x$.

Hence $a = x + b$

and $a = x + c$.

Hence $x + b = x + c$.

Therfor $b = c$. § 49.

115. Theorem. *If $a > b$ and $b > c$, then $a > c$ and $a - c = (a - b) + (b - c)$.*

Since $a > b$, $a = x + b$ and $a - b = x$.

Since $b > c$, $b = y + c$ and $b - c = y$.

Hence $x + b = x + (y + c)$. § 49.

Hence $a = (x + y) + c$ and $a - c = x + y$.

Therfor $a > c$ and $a - c = (a - b) + (b - c)$. § 50.

116. Theorem. *If we hav a series of inequalities of the same kind, as $a > b > c > d > e > \cdots$, any number in the series is greater than any following number.*

The proof of this theorem is identical in form with that of § 36.

117. Theorem. *If we hav a series of equalities or inequalities of the form* $a \gtreqless b \gtreqless c \gtreqless d \gtreqless e \gtreqless \cdots$, *any number in the series is equal to or greater than any following number, the sign being* $>$, *if ther is at least one sign of inequality in the series between the two numbers.*

The proof is like that of § 116.

118. Theorem. *If* $a > b$ *and* $b > c$, *then* $a - c > b - c$ *and* $(a - c) - (b - c) = a - b$; *and conversely, if* $a > c$, $b > c$, *and* $a - c > b - c$, *then* $a > b$ *and* $a - b = (a - c) - (b - c)$.

For the direct theorem,

since $a > b > c$, $a > c$ and $a - c = (a - b) + (b - c)$. § 115.

Therfor $a - c > b - c$

and $(a - c) - (b - c) = a - b$. §§ 75, 79.

For the converse theorem,

since $a > c$, $a = x + c$ and $a - c = x$.

Since $b > c$, $b = y + c$ and $b - c = y$.

Since $a - c > b - c$, $x > y$. § 117.

Hence $x + c > y + c$. § 111.

Or $a > b$. § 117.

Therfor, by the first part of the theorem, since $a > b > c$, $a - b = (a - c) - (b - c)$.

119. Theorem. *If* $a > b$ *and* $b > c$, *then* $a - c > a - b$ *and* $(a - c) - (a - b) = b - c$; *and* *conversely, if* $a > c$, $a > b$, *and* $a - c > a - b$, *then* $b > c$ *and* $b - c = (a - c) - (a - b)$.

120. Theorem. *If* $a > b$ *and* $c = d$, *then* $a + c > b + d$ *and* $(a + c) - (b + d) = a - b$; *and conversely, if* $a + c > b + d$ *and* $c = d$, *then* $a > b$ *and* $a - b = (a + c) - (b + d)$.

For the direct theorem,

since $a > b$, $a + c > b + c$

and $(a + c) - (b + c) = a - b$. § 111.

Since $c = d$, $b + c = b + d$. § 49.

Therfor $a + c > b + d$

and $(a + c) - (b + c) = (a + c) - (b + d)$. § 114.

Also $(a + c) - (b + d) = a - b$. § 34.

For the converse theorem,

since $c = d$, $b + c = b + d$.

Hence, since $a + c > b + d$,

$$a + c > b + c$$

and $(a + c) - (b + d) = (a + c) - (b + c)$.

Therfor $a > b$ and $a - b = (a + c) - (b + c)$.

Also $a - b = (a + c) - (b + d)$.

121. Theorem. *If* $a > b$ *and* $c > d$, *then* $a + c > b + d$ *and* $(a + c) - (b + d) = (a - b) + (c - d)$.

Since $a > b$,

$a + c > b + c$ and $(a + c) - (b + c) = a - b$. § 111.

Since $c > d$,

$b + c > b + d$ and $(b + c) - (b + d) = c - d$. § 112.

Therfor $a + c > b + d$ and $(a + c) - (b + d) =$

$$[(a + c) - (b + c)] + [(b + c) - (b + d)]$$
$$= (a - b) + (c - d). \qquad §§\ 115,\ 50.$$

122. Theorem. *If* $a = b$, $c = d$, *and* $b > c$, *then* $a > c$, $b > d$, *and* $a - c = b - d$; *and conversely, if* $a > c$, $b > d$, $a - c = b - d$, *and* $c = d$, *then* $a = b$; *if* $a > c$, $b > d$, $a - c = b - d$, *and* $a = b$, *then* $c = d$.

For the direct theorem,

since $a = b > c$, $a > c$ and $a - c = b - c$.

Since $b > c = d$, $b > d$ and $b - c = b - d$.

Therfor $a - c = b - d$.

For the first converse theorem,

since $b > d = c$, $b - d = b - c$.

Hence, since $a - c = b - d$,

$a - c = b - c$.

Therfor $a = b$.

Similarly the remaining theorem may be proved.

123. Theorem. *If* $a > b$, $c = d$, *and* $b > c$, *then* $a > c$, $b > d$, $a - c > b - d$, *and* $(a - c) - (b - d) = a - b$; *and conversely, if* $a > c$, $b > d$, $a - c > b - d$, *and* $c = d$, *then* $a > b$.

For the direct theorem,

since $a > b > c$, $\qquad a > c$,

$a - c > b - c$, and $(a - c) - (b - c) = a - b$. § 118.

Since $b > c = d$, $\qquad b > d$

and $\qquad b - c = b - d$.

Therfor $a - c > b - d$

and $\qquad (a - c) - (b - c) = (a - c) - (b - d)$.

Also $\qquad (a - c) - (b - d) = a - b$.

For the converse theorem,

since $b > d = c$, $\qquad b - d = b - c$.

Hence, since $\qquad a - c > b - d$,

$$a - c > b - c.$$

Therfor $\qquad a > b$ § 118.

124. Theorem. *If $a = b$, $c > d$, and $b > c$,* then *$a > c$, $b > d$, $a - c < b - d$, and $(b - d) - (a - c) = c - d$; and conversely, if $a > c$, $b > d$, $a - c < b - d$, and $a = b$, then $c > d$.*

For the direct theorem,

since $a = b > c$, $\qquad a > c$

and $a - c = b - c$.

Since $b > c > d$, $\qquad b > d$,

$b - d > b - c$, and $(b - d) - (b - c) = c - d$.

Therfor $a - c < b - d$

and $$(b - d) - (a - c) = (b - d) - (b - c).$$

Also $$(b - d) - (a - c) = c - d.$$

For the converse theorem,

since $b = a > c$, $\quad b - c = a - c.$

Hence, since $\quad a - c < b - d,$

$$b - c < b - d. \qquad §§ 89, 114.$$

Therfor $$c > d. \qquad § 119.$$

125. Theorem. *If $a > b$, $c > d$, and $b > c$, then $a > d$, $a - d > b - c$, and $(a - d) - (b - c) = (a - b) + (c - d)$.*

Since $b > c > d$, $\quad b > d,$

$$b - d > b - c, \quad \text{and} \quad (b - d) - (b - c) = c - d.$$

Since $a > b > d$, $\quad a > d,$

$$a - d > b - d, \quad \text{and} \quad (a - d) - (b - d) = a - b.$$

Therfor $a - d > b - c$ and $(a - d) - (b - c) =$

$$[(a - d) - (b - d)] + [(b - d) - (b - c)]$$

$$= (a - b) + (c - d). \quad §§ 115, 50.$$

126. Theorem. *If $a > b$, $c > d$, and $a - b = c - d$, then $a + d = b + c$; and conversely, if $a > b$ and $a + d = b + c$, then $c > d$ and $a - b = c - d$.*

For the direct theorem,

since $a > b$, $\quad a + d > b + d$

and $$(a + d) - (b + d) = a - b. \qquad § 111.$$

Since $\qquad c > d,\ b + c > b + d$

and $\qquad (b + c) - (b + d) = c - d.$ § 112.

Hence, since $\qquad a - b = c - d,$

$$(a + d) - (b + d) = (b + c) - (b + d).$$

Therfor $\qquad a + d = b + c.$ § 113.

For the converse theorem,

since $a > b$, $\qquad a + d > b + d$

and $\qquad (a + d) - (b + d) = a - b.$

Hence, since $\qquad a + d = b + c,$

$$b + c > b + d$$

and $(b + c) - (b + d) = (a + d) - (b + d).$ § 113.

Hence $\qquad c > d$

and $\qquad c - d = (b + c) - (b + d).$ § 112.

Therfor $\qquad a - b = c - d.$

127. Definition. From the preceding theorem it follows that if either of the statements "$a > b$, $c > d$, and $a - b = c - d$" and "$a > b$ and $a + d = b + c$" is true, the other is. The truth of either statement is then said to be a **necessary and sufficient condition** for the truth of the other. The truth of either statement follows necessarily from the truth of the other; and the truth of either is sufficient to establish the truth of the other.

Whenever a theorem and its converse ar both true, we have two statements, the hypothesis and conclusion of either theorem, either of which statements is a necessary and sufficient condition for the truth of the other.

128. Theorem. *If* $a > b$, $c > d$, *and* $a - b > c - d$, *then* $a + d > b + c$ *and* $(a + d) - (b + c) = (a - b) - (c - d)$; *and conversely, if* $c > d$ *and* $a + d > b + c$, *then* $a > b$ *and* $a - b > c - d$.

For the direct theorem,

since $a > b$, $\quad a + d > b + d$

and $\quad (a + d) - (b + d) = a - b.$

Since $c > d$, $\quad b + c > b + d$

and $\quad (b + c) - (b + d) = c - d.$

Hence, since $\quad a - b > c - d,$

$$(a + d) - (b + d) > (b + c) - (b + d)$$

and $\quad [(a + d) - (b + d)] - [(b + c) - (b + d)]$
$$= (a - b) - (c - d).$$

Therfor $\quad a + d > b + c$

and $\quad (a + d) - (b + c) = (a - b) - (c - d).$

For the converse theorem,

since $c > d$, $\quad b + c > b + d$

and $\quad (b + c) - (b + d) = c - d.$

Hence, since $$a + d > b + c,$$

$$a + d > b + d$$

and $(a + d) - (b + d) > (b + c) - (b + d)$.

Therfor $a > b$ and $a - b = (a + d) - (b + d)$.

Therfor $$a - b > c - d.$$

129. Theorem. *Having given any two numbers a and b, one of the statements $a = b$, $a > b$, and $a < b$ must be true, and only one can be true* (1).

For either $a = b$ or $a \neq b$.

In the latter case either $a = x + b$ or $x + a = b$. § 45.

Hence either $a > b$ or $a < b$.

Therfor one of the statements $a = b$, $a > b$, and $a < b$ must be true.

Now, if $a = b$, $a \neq x + b$ and $x + a \neq b$. §§ 43, 7, 8.

Hence, if $a = b$, $a \ngtr b$ and $a \nless b$.

If $a > b$, $$a = x + b.$$

Hence $$a \neq b.$$ § 44.

Moreover, if $a > b$, $a \nless b$. For if $a < b$, since also $a > b$, we would hav $a > a$ (§115), which contradicts §105.

Hence, if $a > b$, $a \neq b$ and $a \nless b$.

Similarly it may be shown that if $a < b$, $a \neq b$ and $a \ngtr b$.

---

(1) See Stolz und Gmeiner, p. 6.

4

Therfor only one of the statements $a = b$, $a > b$, and $a < b$ can be true.

This theorem may be more briefly, tho less exactly, stated:

Having given any two numbers a and b, either $a = b$, $a > b$, or $a < b$.

130. Definition. Every proof, when written in its complete form, consists of one or more steps calld **syllogisms**(1). Every syllogism consists of three statements, the **major premis, minor premis,** and **conclusion** (2). The major premis is always a general truth which has already been establisht, either a definition, axiom, or theorem. The minor premis is a particular truth from which and the major premis the conclusion follows as a necessary consequence.

131. Ther ar several kinds of syllogisms, but the ones most frequently used in mathematics ar five in number, two in which the major premis is a categorical statement, two in which the major premis is a hypothetical statement, and one in which it is a disjunctiv statement. These may be represented symbolically as follows:

I. Constructiv Categorical Syllogism.

Major Premis. All B is C.

Minor Premis. A is B.

Conclusion. Therfor A is C.

---

(1) See Jevons-Hill, p. 10.

(2) See Jevons-Hill, p. 113.

II. Destructiv Categorical Syllogism.

Major Premis. All B is C.
Minor Premis. A is not C.
Conclusion. Therfor A is not B.

III. Constructiv Hypothetical Syllogism (1).

Major Premis. If A is B, then C is D.
Minor Premis. A is B.
Conclusion. Therfor C is D.

IV. Destructiv Hypothetical Syllogism.

Major Premis. If A is B, then C is D.
Minor Premis. C is not D.
Conclusion. Therfor A is not B.

V. Disjunctiv Syllogism (2).

Major Premis. Either A is B, or C is D.
Minor Premis. A is not B.
Conclusion. Therfor C is D.

132. Definition. When a constructiv syllogism is used in a proof, the proof is said to be **direct**; when a destructiv syllogism is used, the proof is calld **indirect**, or a **reduction to absurdity** (3); when syllogism V is used, the method of proof is calld the **method of exhaustion** or **exclusion.**

Up to § 129 all our proofs were direct. But in that article we used an indirect method to show that if $a > b$, then $a \not< b$.

(1) See Jevons-Hill, p. 151. (2) See Jevons-Hill, p. 157.

(3) See Beman and Smith, "New Plane and Solid Geometry," 1901, p. 14.

133. Theorem. *In the natural series of numbers each number is one less than the immediately succeeding number.*

For, if $a + 1 = b$, a is one less than b. § 75.

134. Theorem. *In the natural series of numbers each number is less than every succeeding number* (1).

135. Theorem. *Unity is the smallest of all numbers* (2).

For unity is the first number in the natural series and every number has a place in that series.

136. Theorem. *Having given any number, ther is a number greater than it.*

For, if a is any given number, the number $a + 1$ is greater than it.

This theorem is equivalent to the statement that *ther is no greatest number.*

137. Definition. Infinit. Finite. When, having given any number of objects of a particular class, another of the same class can be found, the class is said to hav an **unlimited,** or **infinit** number, of objects; less exactly, the number of objects in that class is said to be unlimited or infinit.

In contrast, the number of objects in a group, which does not hav the above property, is said to be **limited,** or **finite.**

138. Theorem. *The number of numbers is infinit* (3).

---

(1) See Cantor, "Sur les fondements de la Théorie des Ensembles Transfinis," translated from the German by Marotte, 1899, p. 14.

(2) See Cantor, p. 17.

(3) See Lucas, "Théorie des Nombres," 1891, p. 3.

For, having given the numbers 1, 2, 3, ⋯, a, ther is a number $a + 1$ greater than all of these (§ 134) and therfor different from them. § 129.

The natural series of numbers has, then, no limit. It has a beginning, but no end.

139. Theorem. *If $a < b < c$, b is one of the numbers in the natural series between a and c.*

Let $b = a + x$.

The numbers following a in the natural series ar $a + 1, a + 2, a + 3, \cdots$ § 69.

Since x is a number in the series $1, 2, 3, \cdots$ (§ 65), one number of the above series is $a + x$ or b.

Let $c = b + y$.

Then $b + y$ or c is one of the numbers $b + 1, b + 2, b + 3, \cdots$, which follow b in the natural series.

We hav then the portion of the natural series $a, a + 1, a + 2, a + 3, \cdots, b, b + 1, b + 2, b + 3, \cdots, c$.

Therfor b is one of the numbers in the natural series between a and c.

140. Definition. On this account, when $a < b < c$, b is said to **lie between** a and c in magnitude.

141. Theorem. *If a and b ar any two consecutiv numbers in the natural series, ther exists no number between a and b* (1).

For, suppose $a < c < b$.

Then $b - a > b - c \geqq 1$. §§ 119, 135.

Hence $b - a > 1$.

---

(1) See Cantor, p. 15.

But by hypothesis $b = a + 1$.

Hence $b - a = 1$.

Since $b - a$ cannot be both greater than one and equal to one (§ 129), the theorem is establisht.

142. Subtraction. When two unequal numbers a and b ar given, we may determin which is the greater and find their difference, $a - b$ or $b - a$, by the following method.

Consider the two series of numbers

$$b,\ b+1,\ b+2,\ b+3,\ \cdots,\ b+a,\ \cdots$$

$$a,\ a+1,\ a+2,\ a+3,\ \cdots,\ a+b,\ \cdots$$

Now either $a > b$ or $a < b$. § 129.

In the first case $b < a < b + a$. § 108.

In the second $a < b < a + b$.

In the first case, then, a must be one of the numbers in the first series between b and $b + a$. § 139.

In the second case b must be one of the numbers in the second series between a and $a + b$.

In any case, then, either a will be some number $b + x$ in the first series and the difference $a - b$ will be x, or b will be some number $a + y$ in the second series and the difference $b - a$ will be y.

By forming these series, therfor, we may determin whether $a > b$ or $a < b$ and find the difference $a - b$, if $a > b$, or $b - a$, if $a < b$.

143. Thus, to find whether $5 > 2$ or $2 > 5$, and also to find the difference between these numbers, we form the two series

$$5, \ 5 + 1, \ 5 + 2$$

and $$2, \ 2 + 1, \ 2 + 2, \ 2 + 3, \ 2 + 4, \ 2 + 5,$$

or $$5, \ 6, \ 7$$

and $$2, 3, 4, 5, 6, 7.$$

We find that 2 is not in the first series, but 5 is in the second and $5 = 2 + 3$. Hence $5 > 2$ and $5 - 2 = 3$.

Hence also the symbol $2 - 5$ is meaningless. § 87.

144. Theorem. *Any given set of numbers, no two of which ar equal, may be arranged in a series so that each shall be less than the succeeding one.*

Let the given set be $a, b, c, d, e, \cdots$

To arrange these numbers in a series so that each shall be less than the succeeding one, we put down any one number first, say a.

Then we compare any other, as b, with a.

If $b < a$, we set down b in front of a. If $b > a$, we set b down after a.

Suppose $b > a$. Then we hav the series a, b, wher $a < b$.

Next we compare c with a and place it befor or after a according as it is less or greater than a.

Suppose $c > a$. Then we must compare c with b. If $c < b$, we hav the series a, c, b, wher $a < c < b$.

Similarly we may compare each of the other numbers in turn with those already in the series and assign to it a place in the series.

Thus we may hav the series $a, c, b, d, e, \cdots$, wher $a < c < b < d < e < \cdots$

Numbers so arranged ar said to be arranged in **order of magnitude**, or simply **orderd.**

145. Definition. Operation of Subtraction. The process described in § 142 for finding whether $a > b$ or $a < b$ and for finding the difference between a and b, or between b and a is calld **subtraction.**

Thus the difference $a - b$, when it exists, is obtaind from a and b by an operation, the operation of subtraction. In this operation a is the operand, the passiv element, or **minuend,** and b the operator, the activ element, **subtrahend** (1), or **subtrahent** (2).

We may also obtain $a - b$ by the use of objects (3). If from a group of a objects we take away b objects, the number remaining will be $a - b$.

Or, if from a group of a objects we take away enuf objects for the number remaining to be b, the number taken away will be $a - b$.

These results follow directly from § 110.

---

(1) See Schubert, "Encyklopädie," p. 9.

(2) This name, with an activ ending, was proposed by Schröder. See Schubert, "Mathematical Essays," p. 11.

(3) See Tannery, p. 13.

146. Subtraction Table. In order to make a table of the existing differences between all possible pairs of the numbers 1, 2, 3, ···, 10, we divide a square into compartments as in § 74 and write the numbers 1, 2, 3, ···, 10 along the top and left-hand sides of the square as in that article.

In each compartment we wish to place the number which is the difference between the two numbers found respectivly at the left of the row and at the top of the column in which the compartment lies, if this difference exists.

Thus in the sixth row of compartments we should write the numbers that represent the differences $6 - 1$, $6 - 2$, $6 - 3$, ···

To find these differences we may use the addition table (§ 74) to advantage. We notice that in the compartments of that table the 6's lie in a line going from the top downwards and to the left. The 6 of the first row is in the fifth column; hence $6 - 1 = 5$ (§ 79). The 6 of the second row is in the fourth column; hence $6 - 2 = 4$. And so on. Ther is no 6 in any row below the fifth, so that the differences $6 - 6$, $6 - 7$, $6 - 8$, ··· do not exist.

Hence the sixth row of our new table will contain the numbers 5, 4, 3, 2, 1 in the first five compartments and none in the others.

In a similar manner we may form the rest of the table, which is given below.

SUBTRACTION TABLE.

	1	2	3	4	5	6	7	8	9	ϒ	∩	ℭ	ƨ	¥	Ɔ	10
1																
2	1															
3	2	1														
4	3	2	1													
5	4	3	2	1												
6	5	4	3	2	1											
7	6	5	4	3	2	1										
8	7	6	5	4	3	2	1									
9	8	7	6	5	4	3	2	1								
ϒ	9	8	7	6	5	4	3	2	1							
∩	ϒ	9	8	7	6	5	4	3	2	1						
ℭ	∩	ϒ	9	8	7	6	5	4	3	2	1					
ƨ	ℭ	∩	ϒ	9	8	7	6	5	4	3	2	1				
¥	ƨ	ℭ	∩	ϒ	9	8	7	6	5	4	3	2	1			
Ɔ	¥	ƨ	ℭ	∩	ϒ	9	8	7	6	5	4	3	2	1		
10	Ɔ	¥	ƨ	ℭ	∩	ϒ	9	8	7	6	5	4	3	2	1	

147. Definition. An **inverse** of a given operation is an operation that undoes the work of the given operation.

148. Theorem. *The operation of subtraction is inverse to that of addition* ([1]).

For, as we hav seen in § 108, if we add b to a given number and then subtract b, we get back the original number.

149. Theorem. *Addition is inverse to subtraction.*

For (§ 110), if we subtract b from a given number and then add b, we get back the original number.

([1]) See Schubert, "Encyklopädie," p. 8.

150. Definition. The operation of addition is calld a **direct** ([1]), or **synthetic** operation. Having given two numbers we find their sum directly, that is, without the intervention of any other operation.

On the other hand, in subtraction, having given two numbers, of which the second is a part of the first, the object is to find the other part, that is, to find a third number which added to the second makes the first. Every question in subtraction is, then, turnd into a question in addition.

For this reason subtraction is calld an **indirect** or **analytic** operation.

For the same reasons as above given $a+b$ and $a-b$ ar calld respectivly **synthetic** and **analytic combinations** ([2]) of the numbers a and b.

The opposition in nature between the two operations may also be well shown by the equality $c=a+b$.

If a and b ar given to find c, we get the result by addition.

If c and b ar given to find a, or c and a given to find b, we get the result by subtraction, the operations in these two cases being of the same nature because of the commutativ law for addition.

The operation of addition is always possible. The operation of subtraction is possible only when the passiv number is greater than the activ number.

([1]) See Stolz und Gmeiner, p. 5 ; Schubert, Encyklopädie, p. 8.

([2]) See Grassmann, p. 5.

PART II.

INTEGERS.

CHAPTER III.

POSITIV INTEGERS. ZERO. NEGATIV INTEGERS.

151. We now proceed to invent a new system of numbers and will, therfor, for the sake of distinction, call the numbers hitherto treated of **primary numbers** (1).

152. We have seen that, when $a > b$, the symbol $a - b$ has a meaning, that is, the two primary numbers a, b, when written in this order and connected by the sign $-$, determin a certain concept, the difference between a and b. When $a \ngtr b$, the symbol $a - b$ is meaningless and the numbers a, b determin no concept, when written in this order and connected by the sign $-$. It will be useful, however, to hav the numbers a, b determin an idea even when $a \ngtr b$, which shall hav properties like those of the difference $a - b$, when $a > b$.

153. Definition. Let us agree, then, that every two primary numbers a, b, taken in this order, shall determin a concept (2), to be calld an **integer**, this con-

(1) See Lefevre, p. 19.

(2) For this method of defining an integer see Dini, "Grundlagen für eine Theorie der Functionen einer veränderlichen reellen Grösse," trans-

cept to be the difference $a - b$, when $a > b$ (1), that is, when the symbol $a - b$ has a meaning, and in all other cases to hav properties like those of a difference. Let us also call this new concept a number.

Stated briefly, then, an integer is a number couple made by combining two primary numbers a, b in this order, the number couple being the difference $a - b$, when $a > b$, and in all cases having properties like those of a difference.

Thus every two primary numbers determin an integer.

In the case when the integer is a difference it is univalent by § 77; let us agree that in all other cases, also, the two primary numbers shall determin only one integer.

Then every two primary numbers will determin one, and only one, integer.

This concept, the integer determind by the primary numbers a, b, is not the difference between a and b, except when $a > b$, but a new idea created by our imagination, a compound idea made up of the two simple ideas a and b.

154. We might use for this compound idea any symbol to which a meaning has not already been

---

lated from the Italian by Lüroth und Schepp, 1892, p. 3. The idea of "number couple" is due to Wm. Rowan Hamilton. See the *Transactions of the Royal Irish Academy*, 1835, Vol. XVII., part 2, p. 293; "Hamilton's Lectures on Quaternions," 1853, preface, p. 8; Stolz und Gmeiner, p. 51.

(1) See Dini, p. 4; Stolz und Gmeiner, p. 48.

attacht and which should show distinctly the two primary numbers of which the integer is composed. Thus we might use the symbol $a - b$ (1), which by agreement represents the same idea as the integer when $a > b$. But we would then be likely to confuse the idea of integer with the idea of difference. We will therfor use the symbol (a, b) (2).

Thus, for example, (5, 2), (4, 4), and (3, 7) ar integers.

155. Definition. The primary numbers a and b will be calld the **elements** of the integer (a, b), a the **first element** and b the **second.**

156. The order of the elements is essential. Thus (a, b) is not necessarily the same integer as (b, a). In case, however, the two elements ar the same, interchanging them does not alter the integer.

157. For the sake of clearness, we will continue to use the Roman letters $a, b, c, d, \cdots$ to denote primary numbers and will use the Greek letters $\alpha, \beta, \gamma, \delta, \cdots$ to represent integers. We will also employ one Roman letter with the distinguishing **subscripts** 1, 2 to denote the two elements of any given integer, and the corresponding Greek letter to represent the integer itself.

Thus α represents the integer (a_1, a_2), β the integer (b_1, b_2), γ the integer (c_1, c_2), and so on.

---

(1) See statement of Principle of Continuity, or Permanence, in Peacock's "Symbolical Algebra," 1845, p. 2 and in Hankel's "Theorie der komplexen Zahlsysteme," 1867, § 3. See also Lefevre, p. 76.

(2) See Stolz und Gmeiner, p. 47.

158. Theorem. *The system of integers includes the system of primary numbers.*

Let a be any given primary number. Choosing a_2 at random and setting $a_1 = a_2 + a$, we hav $a_1 - a_2 = a$. Hence the integer (a_1, a_2) is the primary number $a_1 - a_2$, or a. § 153.

159. Definition. Integers may be divided into three classes according as the first element is greater than, equal to, or less than the second.

Integers of the first class are calld **positiv,** those of the second **zero,** and those of the third **negativ** (1).

Thus (2, 1), (5, 3), (9, 5) ar positiv integers, (1, 1), (3, 3), (7, 7) ar zero integers, and (1, 3), (2, 7), (4, 9) ar negativ integers.

160. Theorem. *All positiv integers ar primary numbers, and conversely.*

161. Definition. The integer obtaind from a given integer α by interchanging its elements is calld the **opposit,** or **negativ,** of α and is indicated by the symbol $\bar{\alpha}$ or ${}^{-}\alpha$, the sign — being written directly over α, or above and slightly to the left (2). This symbol is univalent (§ 153). Thus, if α is (a_1, a_2), $\bar{\alpha}$ is (a_2, a_1). For example, if α is (8, 3), $\bar{\alpha}$ is (3, 8).

The minus sign, as here used, is a sign of operation.

Sometimes, to more clearly distinguish α from $\bar{\alpha}$, a plus sign is written over α, thus $\overset{+}{\alpha}$, or above and slightly to the left, thus ${}^{+}\alpha$.

---

(1) See Dini, p. 4.

(2) For this symbol see Stolz und Gmeiner, p. 54.

Since every primary number is an integer (§ 160), every primary number has its opposit.

162. Theorem. *Every negativ integer is the opposit of some primary number.*

Thus, suppose (a_1, a_2) is negativ.

Then $a_1 < a_2$. § 159.

Now (a_1, a_2) is the opposit of (a_2, a_1) wher $a_2 > a_1$.

Hence (a_1, a_2) is the opposit of the primary number $a_2 - a_1$, that is, (a_1, a_2) is the same as $^-(a_2 - a_1)$.

If $a_2 - a_1 = a$, (a_1, a_2) is the same as $\overline{a}$.

For example, (4, 6) is $\overline{2}$, (5, 9) is $\overline{4}$, (1, 8) is $\overline{7}$.

163. The signs + and — over primary numbers may be regarded as signs of quality showing whether the integer thus represented is positiv or negativ.

164. Theorem. *The opposit of a positiv integer is a negativ integer; the opposit of a zero integer is a zero integer; the opposit of a negativ integer is a positiv integer.*

If (a_1, a_2) is positiv, $a_1 > a_2$.

Hence $a_2 < a_1$ and (a_2, a_1) is negativ.

Similarly the other parts of the theorem may be proved.

165. Theorem. *The opposit of the opposit of a given integer is the given integer.*

166. Theorem. *If we interchange the elements of an integer an odd number of times, the resulting integer is the opposit of the given integer; if we interchange the elements an even number of times, the result is the given integer.*

167. Let us proceed now to giv integers properties analogous to those possest by differences. Let us define the terms equal, greater, less, sum, difference, etc., in connection with integers. These terms ar at present meaningless with reference to integers. We can therfor giv any definitions for them that we please, provided the definitions ar not inconsistent with each other or with preceding definitions. We must remember in particular that, when $a > b$, the integer (a, b) is the difference $a - b$, a primary number, and that for such integers the above terms ar already defined.

168. Definition. Equality (1). To obtain a proper definition for the term "equality" with reference to two integers (a_1, a_2) and (b_1, b_2), let us consider first the case when $a_1 > a_2$ and $b_1 > b_2$, in which case these integers ar the primary numbers $a_1 - a_2$ and $b_1 - b_2$.

By § 126 a necessary and sufficient condition for the equality of $a_1 - a_2$ and $b_1 - b_2$, when $a_1 > a_2$ and $b_1 > b_2$, is that $a_1 + b_2$ should equal $a_2 + b_1$. If we make this condition our requirement for the equality of any two integers (a_1, a_2) and (b_1, b_2), ther will be no contradiction of definitions and the property of equality of two integers will be like that of equality of differences. Formally then we define: $(a_1, a_2) = (b_1, b_2)$, if $a_1 + b_2 = a_2 + b_1$.

Thus $(7, 4) = (9, 6)$, since $7 + 6 = 4 + 9$; $(5, 9) = (1, 5)$, since $5 + 5 = 9 + 1$.

(1) See Dini, p. 3; Davis, "Logic of Algebra," 1890, p. 17; Stolz und Gmeiner, p. 48.

169. Theorem. *If* $a_1 = b_1$ *and* $a_2 = b_2$, *then* $(a_1, a_2) = (b_1, b_2)$.

For, then, $a_1 + b_2 = a_2 + b_1$. § 50.

170. Theorem. $\alpha = \alpha$, or $(a_1, a_2) = (a_1, a_2)$.

This follows from § 169,

since $a_1 = a_1$ and $a_2 = a_2$. § 6.

171. Theorem. *If* $\alpha = \beta$, *then* $\beta = \alpha$.

Since $\alpha = \beta$, or $(a_1, a_2) = (b_1, b_2)$,

$a_1 + b_2 = a_2 + b_1$. § 168.

Hence $a_2 + b_1 = a_1 + b_2$. § 7.

Or $b_1 + a_2 = b_2 + a_1$. § 41.

Therfor $\beta = \alpha$. § 168.

This theorem is the **Commutativ Law for Equality of Integers.**

172. Theorem. *If* $\alpha \neq \beta$, *then* $\beta \neq \alpha$.

This is proved in the same way as § 8.

173. Theorem. *If* $\alpha = \beta$ *and* $\beta = \gamma$ *then* $\alpha = \gamma$.

Since $\alpha = \beta$, $a_1 + b_2 = a_2 + b_1$.

Since $\beta = \gamma$, $b_1 + c_2 = b_2 + c_1$.

Adding, $(a_1 + b_2) + (b_1 + c_2) = (a_2 + b_1) + (b_2 + c_1)$. § 50.

Or $(a_1 + c_2) + (b_1 + b_2) = (a_2 + c_1) + (b_1 + b_2)$. § 59.

Hence $a_1 + c_2 = a_2 + c_1$. § 48.

Therfor $\alpha = \gamma$.

174. Theorem. *If* $\alpha = \beta$ *and* $\beta \neq \gamma$, *then* $\alpha \neq \gamma$.

This is proved in the same way as § 35.

175. Theorem. *If we hav a series of equalities in which the right member of each equality, except the last, is also the left member of the succeeding equality, as* $\alpha = \beta = \gamma = \delta = \varepsilon = \cdots$, *then any integer in the series is equal to any other integer in the series.*

176. Theorem. *If* $\alpha = \beta$, *then* α *is positiv if* β *is positiv,* α *is zero if* β *is zero, and* α *is negativ if* β *is negativ.*

Since $\alpha = \beta$, $\qquad a_1 + b_2 = a_2 + b_1$.

If β is positiv, $\qquad b_1 > b_2$.

Hence $\qquad a_1 > a_2$. § 126.

Therfor α is positiv.

Similarly the other parts of the theorem may be proved.

177. Theorem. *If* $\alpha = \beta$, *then* $\bar{\alpha} = \overline{\beta}$.

178. Theorem. *If we hav a series of equal positiv integers, they all represent the same primary number.*

Suppose $\alpha = \beta = \gamma = \cdots$

Then, since $\alpha, \beta, \gamma, \cdots$ ar all positiv, this means that $a_1 - a_2 = b_1 - b_2 = c_1 - c_2 = \cdots$

That is, $a_1 - a_2$, $b_1 - b_2$, $c_1 - c_2$, $\cdots$ ar the same primary number. § 5.

Therfor $\alpha, \beta, \gamma, \cdots$ all represent the same primary number.

Thus (4, 2), (5, 3), (8, 6) all represent the primary number 2.

179. Theorem. *If we hav a series of equal negativ integers, they ar all opposits of the same primary number.*

Suppose $\alpha = \beta = \gamma = \cdots$ and that $\alpha, \beta, \gamma, \cdots$ ar negativ.

Then $\bar{\alpha} = \bar{\beta} = \bar{\gamma} = \cdots$ and $\bar{\alpha}, \bar{\beta}, \bar{\gamma}, \cdots$ ar positiv. §§ 177, 164.

Hence $\bar{\alpha}, \bar{\beta}, \bar{\gamma}, \cdots$ all represent the same primary number. § 178.

But $\alpha, \beta, \gamma, \cdots$ ar the opposits of $\bar{\alpha}, \bar{\beta}, \bar{\gamma}, \cdots$ § 165.

Therfor $\alpha, \beta, \gamma, \cdots$ ar all opposits of the same primary number.

Thus (2, 7), (3, 8), (4, 9) ar all opposits of the primary number 5 and may be represented by the symbol $\bar{5}$.

180. Theorem. *All zero integers ar equal* [1].

Let α and β be any two zero integers.

Then $a_1 = a_2$ and $b_1 = b_2$. § 159.

Hence $a_1 + b_2 = a_2 + b_1$.

Therfor $$\alpha = \beta.$$

Hence, if $\alpha, \beta, \gamma, \cdots$ ar a series of zero integers, $\alpha = \beta = \gamma = \cdots$

181. Definition. We saw in § 178 that a series of equal positiv integers all represent the same idea, a certain primary number.

In §§ 179, 180 we saw that a series of equal negativ integers ar all opposits of a certain primary number and that all zero integers ar equal.

---

(1) See Dini, p. 4.

We will therfor, for the sake of uniformity in our system of numbers, say that all equal negativ integers ar the **same** number and that all zero integers ar the **same** number. The latter number we will call **zero, cipher,** or **naught** and will represent by the symbol o [1].

Thus o stands for (3, 3), (5, 5), (8, 8), ⋯

182. We may therfor henceforth, when any two integers ar equal, say they ar the same integer.

183. Theorem. $\bar{0} = 0$.

184. Theorem. $\bar{\bar{a}} = a$.

185. Theorem. *If the same primary number be added to each element of an integer, the integer is unchanged.*

In symbols $(a_1 + k, a_2 + k) = (a_1, a_2)$.

This follows immediately from the definition of equality.

186. Definition. Greater. Less[2]. In order to discover a proper definition for the terms "greater" and "less" with reference to two integers (a_1, a_2) and (b_1, b_2), let us consider first the case when these integers ar the primary numbers $a_1 - a_2$ and $b_1 - b_2$.

By § 128 a necessary and sufficient condition that $a_1 - a_2$ should be greater than $b_1 - b_2$, when $a_1 > a_2$ and $b_1 > b_2$, is that $a_1 + b_2$ should be greater than $a_2 + b_1$. If we make this condition the meaning of the statement "(a_1, a_2) is greater than (b_1, b_2)," ther will be

(1) The symbol o is used in an Indian manuscript of the eighth century to denote the absence or lack of number.

(2) See Davis, p. 17; Stolz und Gmeiner, p. 55.

no contradiction of definitions. Formally, then, we define : $(a_1, a_2) > (b_1, b_2)$, if $a_1 + b_2 > a_2 + b_1$.

Similarly, we define: $(a_1, a_2) < (b_1, b_2)$, if $a_1 + b_2 < a_2 + b_1$.

The properties of inequality of integers ar, then, like those of inequality of differences.

187. Theorem. *Having given any two integers α and β, one of the statements $\alpha = \beta$, $\alpha > \beta$, and $\alpha < \beta$ must be true, and only one can be true.*

For one of the statements $a_1 + b_2 = a_2 + b_1$, $a_1 + b_2 > a_2 + b_1$, and $a_1 + b_2 < a_2 + b_1$ must be true, and only one can be. § 129.

188. Theorem. *If $\alpha > \beta$, then $\beta < \alpha$, and conversely.*

If $\alpha > \beta$, $a_1 + b_2 > a_2 + b_1$. § 186.

Hence $a_2 + b_1 < a_1 + b_2$. § 89.

Or $b_1 + a_2 < b_2 + a_1$. §§ 41, 117.

That is, $\beta < \alpha$. § 186.

For the converse theorem reverse the steps.

189. Theorem. *If $\alpha > \beta$, then $\bar{\alpha} < \bar{\beta}$.*

190. Theorem. *If $\alpha = \beta$ and $\beta > \gamma$, then $\alpha > \gamma$.*

The proof of this theorem is formally the same as that of § 173.

191. Theorem. *If $\alpha > \beta$ and $\beta > \gamma$, then $\alpha > \gamma$.*

192. Theorem (1). *If $\alpha < \beta$ and $\beta = \gamma$, then $\alpha < \gamma$.*

Since $\alpha < \beta = \gamma$, $\gamma = \beta > \alpha$. §§ 171, 188.

Therfor $\gamma > \alpha$. § 190.

Or $\alpha < \gamma$. § 188.

(1) See Stolz und Gmeiner, p. 6.

193. Theorem. *If $\alpha < \beta$ and $\beta < \gamma$, then $\alpha < \gamma$.*

The proof of this theorem is like that of § 192.

194. Theorem (1). *If $\alpha > \beta$ and $\beta = \gamma$, then $\alpha > \gamma$.*

For either $\alpha = \gamma$, $\alpha < \gamma$, or $\alpha > \gamma$. § 187.

If $\alpha = \gamma$, since also $\gamma = \beta$, $\alpha = \beta$. § 173.

But this contradicts the hypothesis. § 187.

Hence $\alpha \neq \gamma$

Similarly $\alpha \nless \gamma$.

Therfor $\alpha > \gamma$.

195. Theorem. *If $\alpha = \beta$ and $\beta < \gamma$, then $\alpha < \gamma$.*

This follows from § 194 by the method used to prove § 192.

196. Theorem. *If we hav a series of inequalities of the same kind, as $\alpha > \beta > \gamma > \delta > \varepsilon > \cdots$, any integer in the series is greater than any following integer.*

The proof is the same as that of § 116.

197. Theorem. *If we hav a series of equalities or inequalities of the form $\alpha \gtreqless \beta \gtreqless \gamma \gtreqless \delta \gtreqless \varepsilon \gtreqless \cdots$, any integer in the series is equal to or greater than any following integer, the sign being $>$, if ther is at least one sign of inequality in the series between the two integers.*

The proof is the same as that of § 117.

198. Theorem. *Any given set of integers, no two of which ar equal, may be arranged in order of magnitude.*

This theorem is proved in the same way as § 144.

199. Theorem. *In any given set of integers, no two of which ar equal, ther is one greatest and one least.*

(1) See Stolz und Gmeiner, p. 6.

When the given set is arranged in order of magnitude, the last integer is greater than all the others (§ 196); the first integer is smaller than all the others.

Ther cannot be two, each of which is greater than all the others.

For, suppose λ and μ were each greater than all the others.

Then $\lambda > \mu$ and $\mu > \lambda$.

Since $\mu > \lambda$, $\lambda < \mu$.

But this contradicts § 187.

Hence ther is only one integer greater than all the others.

That is, ther is one greatest integer.

Similarly ther is one least integer.

200. Theorem. *Every positiv integer is greater than zero; every negativ integer is less than zero; and conversely, every integer greater than zero is positiv; every integer less than zero is negativ.*

For, if (a_1, a_2) is positiv, $a_1 > a_2$.

Hence $a_1 + b > a_2 + b$. § 111.

Therfor $(a_1, a_2) > (b, b)$.

Or $(a_1, a_2) > 0$.

Similarly the second part of the theorem is proved.

For the converse theorem reverse the steps.

201. Theorem. *Every positiv integer is greater than every negativ integer; every integer greater than a positiv integer is positiv; every integer less than a negativ integer is negativ.*

202. We saw in §§ 5, 75 how to compare two positiv integers; in § 201 we learnd that every positiv integer is greater than every negativ integer. The following theorem will giv a rule for comparing two negativ integers.

Theorem. *If* $\bar{a} > \bar{b}$, $a < b$; *if* $\bar{a} = \bar{b}$, $a = b$; *if* $\bar{a} < \bar{b}$, $a > b$; *and conversely, if* $a < b$, $\bar{a} > \bar{b}$; *if* $a = b$, $\bar{a} = \bar{b}$; *if* $a > b$, $\bar{a} < \bar{b}$.

This follows immediately from §§ 177, 189.

This theorem might be abbreviated as follows:

$$\bar{a} \gtreqless \bar{b} \text{ according as } a \lesseqgtr b.$$

203. Definition. Sum (1). In order to derive a proper definition for the sum of two integers (a_1, a_2) and (b_1, b_2), let us consider first the case when these integers ar the primary numbers $a_1 - a_2$ and $b_1 - b_2$.

By § 121 the sum of the primary numbers $a_1 - a_2$ and $b_1 - b_2$ is $(a_1 + b_1) - (a_2 + b_2)$, which with our present notation is represented by $(a_1 + b_1, a_2 + b_2)$. Moreover the symbol $(a_1 + b_1, a_2 + b_2)$ is univalent in all cases, since $a_1 + b_1$ and $a_2 + b_2$ ar univalent. §§ 40, 153.

If, then, we make the integer $(a_1 + b_1, a_2 + b_2)$ the sum of (a_1, a_2) and (b_1, b_2) in all cases, ther will be no contradiction of definitions and the sum of two integers will hav a form similar to that of the sum of two differences.

(1) See Dini, p. 3.

Formally, then, we define: $(a_1, a_2) + (b_1, b_2)$ represents $(a_1 + b_1, a_2 + b_2)$.

204. Operation of Addition of Integers. Since in getting the sum of two integers (a_1, a_2) and (b_1, b_2) we add a_1 to b_1 and a_2 to b_2, we may be said to obtain that sum by an operation, the operation of **adding** (a_1, a_2) and (b_1, b_2).

205. Theorem. $(a_1, a_2) + (b_1, b_2) = (a_1 + b_1, a_2 + b_2)$.

For, by definition, $(a_1, a_2) + (b_1, b_2)$ stands for $(a_1 + b_1, a_2 + b_2)$.

And any integer is equal to itself. § 170.

206. Theorem. *The sum of two positiv integers is a positiv integer; the sum of two zero integers is a zero integer; the sum of two negativ integers is a negativ integer; the sum of a positiv and a zero integer is positiv; the sum of a negativ and a zero integer is negativ; and conversely, if the sum of two integers is positiv, one of them is positiv; if the sum is negativ, one is negativ.*

Suppose α and β ar both positiv.

Then $a_1 > a_2$ and $b_1 > b_2$. § 159.

Hence $a_1 + b_1 > a_2 + b_2$. § 121.

Therfor $\alpha + \beta$ is positiv.

Similarly the other parts of the direct theorem may be proved.

The converse theorem is proved by the method of exhaustion.

207. Theorem. $\alpha + \beta = \beta + \alpha$.

For $a_1 + b_1 = b_1 + a_1$ and $a_2 + b_2 = b_2 + a_2$. § 41.

This theorem is the **Commutativ Law for Addition of Integers.**

208. Theorem. $\alpha + \bar{\alpha} = 0$.

For $(a_1, a_2) + (a_2, a_1) = (a_1 + a_2, a_2 + a_1)$, which equals zero. § 159.

209. Theorem. $\alpha + 0 = \alpha$.

For $(a_1, a_2) + (b, b) = (a_1 + b, a_2 + b)$, which equals (a_1, a_2). § 185.

210. Theorem. $0 + 0 = 0$.

211. Theorem. $^-(\alpha + \beta) = \bar{\alpha} + \bar{\beta}$.

In words: *The opposit of the sum of two integers is the sum of their opposits.*

212. Theorem. *If* $\alpha = \beta$, *then* $\alpha + \gamma = \beta + \gamma$, *and conversely.*

For the direct theorem,

since $\alpha = \beta$, $a_1 + b_2 = a_2 + b_1$. § 168.

Adding $c_1 + c_2$ to each member of this equality,

$$(a_1 + b_2) + (c_1 + c_2) = (a_2 + b_1) + (c_1 + c_2). \text{ § 46.}$$

Or $(a_1 + c_1) + (b_2 + c_2) = (a_2 + c_2) + (b_1 + c_1)$. § 59.

Therfor $(a_1 + c_1, a_2 + c_2) = (b_1 + c_1, b_2 + c_2)$. § 168.

Or $\alpha + \gamma = \beta + \gamma$. § 203.

For the converse theorem reverse the steps.

213. Theorem. *If* $\alpha = \beta$, *then* $\gamma + \alpha = \gamma + \beta$, *and conversely.*

This follows from §§ 207, 212, or it may be proved independently of §§ 207, 212 by a method similar to that used for § 212.

214. Theorem. *If* $\alpha = \beta$ *and* $\gamma = \delta$, *then* $\alpha + \gamma = \beta + \delta$; *and conversely, if* $\alpha + \gamma = \beta + \delta$ *and* $\gamma = \delta$, *then* $\alpha = \beta$.

This is proved in the same way as § 50.

215. Theorem. *If* $\alpha > \beta$, *then* $\alpha + \gamma > \beta + \gamma$, and conversely.

The proof of this theorem is formally the same as that of § 212.

216. Theorem. *If* $\alpha > \beta$, *then* $\gamma + \alpha > \gamma + \beta$, *and conversely.*

217. Theorem. *If* $\alpha > \beta$ *and* $\gamma = \delta$, *then* $\alpha + \gamma > \beta + \delta$; *and conversely, if* $\alpha + \gamma > \beta + \delta$ *and* $\gamma = \delta$, *then* $\alpha > \beta$.

218. Theorem. *If* $\alpha > \beta$ *and* $\gamma > \delta$, *then* $\alpha + \gamma > \beta + \delta$.

219. Theorem. *If* $\alpha \gtreqless \beta$ *and* $\gamma \gtreqless \delta$, *then* $\alpha + \gamma \gtreqless \beta + \delta$.

220. Definition. Difference. When (a_1, a_2) and (b_1, b_2) ar the primary numbers $a_1 - a_2$ and $b_1 - b_2$ respectivly, and $a_1 - a_2 > b_1 - b_2$, their difference is, by § 128, $(a_1 + b_2) - (a_2 + b_1)$, which with our present notation is represented by $(a_1 + b_2, a_2 + b_1)$. Moreover the symbol $(a_1 + b_2, a_2 + b_1)$ is univalent in all cases, since $a_1 + b_2$ and $a_2 + b_1$ ar univalent. To avoid contradictions we will make the integer $(a_1 + b_2, a_2 + b_1)$ the difference of (a_1, a_2) and (b_1, b_2) in all cases. Thus the difference of two integers is in form similar to the difference of two differences.

Formally then: $(a_1, a_2) - (b_1, b_2)$ represents $(a_1 + b_2, a_2 + b_1)$.

221. Operation of Subtraction of Integers. The difference $\alpha - \beta$ of two integers α and β may be said to be obtaind by an operation, the operation of **subtracting** β from α.

The definition of § 220 removes the restriction that in subtraction the minuend must be greater than the subtrahend. The operation of subtraction is now possible in all cases.

222. Theorem. $(a_1, a_2) - (b_1, b_2) = (a_1 + b_2, a_2 + b_1)$.

223. Theorem. $\alpha - \beta = \alpha + \bar{\beta}$.

224. Definition. For this reason $\alpha - \beta$ is sometimes calld an **algebraic sum.**

225. Theorem. $\alpha - \bar{\beta} = \alpha + \beta$.

226. Definition. A plus or minus sign directly befor an integer and not following another integer is at present meaningless. Let us agree that such a sign shall be equivalent to a plus or minus sign over the integer.

Thus $+ \alpha + \beta$ is the same as $\overset{+}{\alpha} + \beta$ or $\alpha + \beta$;

$- \alpha + \beta$ is the same as $\bar{\alpha} + \beta$.

227. Theorem. From §§ 184, 223, 225, 226 it follows that *whenever we hav the sign + or — over an integer and the sign + or — befor it, we may change either of these signs, provided the other is also changed from + to — or from — to +.*

Thus $+\overset{+}{a}$ may be changed to $-\overline{a}$;
$+\overline{a}$ may be changed to $-\overset{+}{a}$;
$-\overset{+}{a}$ may be changed to $+\overline{a}$;
$-\overline{a}$ may be changed to $+\overset{+}{a}$.

228. Theorem. $\alpha - \alpha = 0$.

229. Theorem. $\alpha - 0 = \alpha$.

230. Theorem. $0 - \alpha = \overline{\alpha}$.

231. Theorem. $0 - 0 = 0$.

232. Theorem. $-(\alpha - \beta) = -\alpha + \beta$.

233. Theorem. *If $\alpha = \beta$, $\alpha - \beta$ is zero, and conversely.*

First, since $\alpha = \beta$, $a_1 + b_2 = a_2 + b_1$. § 168.
Hence $(a_1 + b_2, a_2 + b_1)$ is zero. § 159.
That is, $\alpha - \beta$ is zero. § 220.
For the converse theorem reverse the steps.

234. Theorem. *If $\alpha > \beta$, $\alpha - \beta$ is positiv; and conversely.*

The proof is formally the same as that of § 233.

235. Theorem. *If $\alpha < \beta$, $\alpha - \beta$ is negativ; and conversely.*

236. Definition. If $\alpha - \beta$ is equal to the positiv integer γ, α is said to be γ **greater** than β, and β is said to be γ **less** than α.

237. Theorem. *If $\alpha = \beta$, then $\alpha - \gamma = \beta - \gamma$, and conversely.*

For, if $\alpha = \beta$, $\quad \alpha + \overline{\gamma} = \beta + \overline{\gamma}$. § 212.

Or $\quad \alpha - \gamma = \beta - \gamma$. § 223.

For the converse theorem reverse the steps.

238. Theorem. *If* $\alpha = \beta$, *then* $\gamma - \alpha = \gamma - \beta$, *and conversely.*

239. Theorem. *If* $\alpha = \beta$ *and* $\gamma = \delta$, *then* $\alpha - \gamma = \beta - \delta$; *and conversely, if* $\alpha - \gamma = \beta - \delta$ *and* $\gamma = \delta$, *then* $\alpha = \beta$; *if* $\alpha - \gamma = \beta - \delta$ *and* $\alpha = \beta$, *then* $\gamma = \delta$.

The first part of this theorem is usually stated: *Equals subtracted from equals leave equals.*

240. Theorem. $(a_1, a_2) = a_1 - a_2$ *in all cases.*

For $a_1 = (a_1 + b, b)$ § 153.

and $a_2 = (a_2 + c, c)$.

Hence $a_1 - a_2 = ((a_1 + b) + c, b + (a_2 + c))$ § 239.

$= (a_1 + (b + c), a_2 + (b + c))$ §§ 54, 169.

$= (a_1, a_2)$. § 185.

Therfor $(a_1, a_2) = a_1 - a_2$. §§ 175, 171.

241. Theorem. *If* $\alpha > \beta$, *then* $\alpha - \gamma > \beta - \gamma$, *and conversely.*

The proof of this is formally the same as that of § 237.

242. Theorem. *If* $\alpha > \beta$, *then* $\gamma - \alpha < \gamma - \beta$, *and conversely.*

Since $\alpha > \beta$, $\bar{\alpha} < \bar{\beta}$. § 189.

Hence $\gamma + \bar{\alpha} < \gamma + \bar{\beta}$. §§ 188, 216.

Or $\gamma - \alpha < \gamma - \beta$. §§ 223, 197.

243. Theorem. *If* $\alpha > \beta$ *and* $\gamma = \delta$, *then* $\alpha - \gamma > \beta - \delta$; *and conversely, if* $\alpha - \gamma > \beta - \delta$ *and* $\gamma = \delta$, *then* $\alpha > \beta$.

244. Theorem. *If* $\alpha = \beta$ *and* $\gamma > \delta$, *then* $\alpha - \gamma < \beta - \delta$; *and conversely, if* $\alpha - \gamma < \beta - \delta$ *and* $\alpha = \beta$, *then* $\gamma > \delta$.

245. Theorem. *If* $\alpha > \beta$ *and* $\gamma < \delta$, *then* $\alpha - \gamma > \beta - \delta$.

246. Theorem. *If* $\alpha \gtreqless \beta$ *and* $\gamma \lesseqgtr \delta$, *then* $\alpha - \gamma \gtreqless \beta - \delta$.

247. Definition. As we connected sums and differences of primary numbers by plus and minus signs, so we may connect sums and differences of integers by plus and minus signs, using parentheses to indicate the order in which the various operations of addition and subtraction ar performd. Every such expression is univalent, since every sum and every difference is univalent. §§ 203, 220.

Such an expression is calld a **complex algebraic sum.** The integers from which it is formd ar calld its **elements.**

If the signs between integers in a complex algebraic sum ar all plus, the complex algebraic sum is calld simply a **complex sum.**

248. Theorem. *If for one or more elements of a complex algebraic sum equal integers ar substituted, the complex sum is unchanged.*

249. Theorem. $(\alpha + \beta) + \gamma = \alpha + (\beta + \gamma)$.

For $\quad (\alpha + \beta) + \gamma = ((a_1 + b_1) + c_1, (a_2 + b_2) + c_2)$

and $\quad \alpha + (\beta + \gamma) = (a_1 + (b_1 + c_1), a_2 + (b_2 + c_2))$.

But $\quad (a_1 + b_1) + c_1 = a_1 + (b_1 + c_1)$ § 54.

and $\quad (a_2 + b_2) + c_2 = a_2 + (b_2 + c_2)$.

Therfor $(\alpha + \beta) + \gamma = \alpha + (\beta + \gamma)$.

This theorem is the **Associativ law for Addition of Integers.**

250. Theorem. *Any complex sum of integers is independent of the arrangement of its parentheses and the order of its elements.*

This is proved in the same way as the corresponding theorem for primary numbers, § 59.

This theorem is the **Generalized Associativ and Commutativ Law for Addition of Integers.**

251. Hence we may speak of a sum $\alpha + \beta + \gamma + \delta + \varepsilon + \cdots$ without any parentheses, and may write the elements of the sum, $\alpha, \beta, \gamma, \delta, \varepsilon, \cdots$, in any order we please without altering it.

252. Theorem. *The sum of the left members of any number of equalities is equal to the sum of the right members.*

Thus, if $\alpha = \beta$, $\gamma = \delta$, $\varepsilon = \zeta$, $\eta = \theta$, $\cdots$, then

$$\alpha + \gamma + \varepsilon + \eta + \cdots = \beta + \delta + \zeta + \theta + \cdots$$

This may be proved in the same way as § 57.

253. Theorem. *If* $\alpha \gtreqless \beta$, $\gamma \gtreqless \delta$, $\varepsilon \gtreqless \zeta$, $\eta \gtreqless \theta$, $\cdots$, *then* $\alpha + \gamma + \varepsilon + \eta + \cdots \gtreqless \beta + \delta + \zeta + \theta + \cdots$

If the sign of one of the given statements is $>$, the sign of the result is $>$.

254. Theorem. *The sum of any number of positiv integers is a positiv integer; the sum of any number of zero integers is a zero integer; the sum of any number of negativ integers is a negativ integer.*

To prove the first part of the theorem, we know that the sum of two positiv integers is a positiv integer (§ 206). Hence the sum of that sum and a positiv integer is positiv, and so on. By applying the rule $n - 1$ times we prove that the sum of n positiv integers is a positiv integer.

Similarly the other parts of the theorem ar proved.

255. Theorem. *The sum of any number of zeros is equal to zero.*

256. Theorem. $-(\alpha + \beta + \cdots + \zeta + \eta + \theta)$

$$= \overline{\alpha} + \overline{\beta} + \cdots + \overline{\zeta} + \overline{\eta} + \overline{\theta}.$$

For

$$-(\alpha + \beta + \cdots + \zeta + \eta + \theta)$$

$$= -\{([(\alpha + \beta + \cdots) + \zeta] + \eta) + \theta\}$$

$$= -([(\alpha + \beta + \cdots) + \zeta] + \eta) + \overline{\theta}$$

$$= -[(\alpha + \beta + \cdots) + \zeta] + \overline{\eta} + \overline{\theta}$$

$$= \text{etc.}$$

257. Consider next any complex algebraic sum. Change all the $-$ signs befor integers and befor parentheses to $+$, at the same time changing the signs over the corresponding integers or parentheses (§ 227). Next remove all $-$ signs from over parentheses, changing the signs over all the integers in these parentheses (§ 256).

Take the following expression as an example:

$$-\{[\alpha + (-\beta - ((\gamma - \overline{\delta}) + (\overline{\varepsilon} - \zeta)) + \eta) - \theta] + (\delta + \beta)\}.$$

Making the above mentiond changes, this becomes

$$\{[\overline{\alpha} + (\beta + ((\gamma + \delta) + (\overline{\varepsilon} + \overline{\zeta})) + \overline{\eta}) + \theta] + (\overline{\delta} + \overline{\beta})\}.$$

We now hav an expression in which all the signs connecting integers ar + signs. All parentheses may therfor be removed (§ 251). Thus we get

$$\bar{\alpha} + \beta + \gamma + \delta + \bar{\varepsilon} + \bar{\zeta} + \bar{\eta} + \theta + \bar{\delta} + \bar{\beta}.$$

The sign over any given integer in the result can be obtaind by counting the minus signs which ar over the integer, befor it, and befor the parentheses within which it lies in the original expression. These ar calld the minus signs that **affect** the given integer.

If the number of these minus signs is odd, the sign over the integer in the result will be minus; if their number is even, the sign over the integer will be plus, or ther will be no sign over it (§ 166). For, every time a minus sign is removed from befor an integer, the sign over the integer must be changed; and the same must be done when a minus sign is removed from in front of a parenthesis in which the integer lies.

Thus in the given expression ther ar four minus signs affecting δ, one over it, one befor it, and two in front of parentheses within which it lies. Therfor the sign of δ in the result is plus. Ther ar three minus signs affecting ε. Hence the sign of ε in the result is minus.

The principle proved above may be stated as follows:

Theorem. *Having given any complex algebraic sum, we may remove all parentheses, if we write a plus sign befor every integer and place over the integer a minus or* *a plus sign, according as the number of minus signs affecting it is odd or even.*

This theorem may be calld the **Associativ Law for Addition and Subtraction of Integers.**

258. Definition. The expression considerd above may be written without any signs over its integers as follows:

$$\{([(\{([(-\alpha+\beta)+\gamma]+\delta)-\varepsilon\} -\zeta)-\eta]+\theta)-\delta\}-\beta. \qquad \S\S\ 250,\ 227.$$

Such an expression is calld a **standard algebraic sum.**

The parentheses ar usually omitted from a standard algebraic sum. Thus the above expression may be written $-\alpha+\beta+\gamma+\delta-\varepsilon-\zeta-\eta+\theta-\delta-\beta$.

In such an expression the operations ar supposed to be performd from left to right ([1]). A parenthesis could not be inserted as follows:

$$-\alpha+\beta+\gamma+\delta-(\varepsilon-\zeta)-\eta+\theta-\delta-\beta.$$

In what follows we will use the term algebraic sum to indicate a standard algebraic sum whose parentheses hav been removed.

259. The integers in the algebraic sum under consideration in § 257 may also be arranged in any order § 250.

([1]) See E. Bardey, "Resultaten zur Aufgabensammlung," 1871, p. 2; Schröder, "Lehrbuch der Arithmetik und Algebra," 1873, Vol. I, p. 219; Stolz und Gmeiner, p. 9.

Thus $\bar{\alpha} + \beta + \gamma + \delta + \bar{\varepsilon} + \bar{\zeta} + \bar{\eta} + \theta + \bar{\delta} + \bar{\beta}$ may be written $\beta + \delta + \bar{\alpha} + \gamma + \bar{\varepsilon} + \bar{\zeta} + \theta + \bar{\eta} + \bar{\beta} + \delta$,
or $\beta + \delta - \alpha + \gamma - \varepsilon - \zeta + \theta - \eta - \beta - \delta$.

We may state this result as follows:

Theorem. *Having given any algebraic sum, the order of its elements may be changed in any way, provided the sign befor each element goes with it.*

This may be calld the **Commutativ Law for Addition and Subtraction of Integers.**

260. A convenient arrangement of the elements in any algebraic sum is one in which all those preceded by the plus sign ar written first and then those preceded by the minus sign.

Thus the expression considerd above would become $\beta + \gamma + \delta + \theta - \alpha - \varepsilon - \zeta - \eta - \delta - \beta$.

The integers preceded by the minus sign may now be enclosed in a parenthesis with plus signs between them and a minus sign befor the parenthesis (§§ 223, 256). Thus the above expression is equal to

$$\beta + \gamma + \delta + \theta - (\alpha + \varepsilon + \zeta + \eta + \delta + \beta).$$

This result may be stated in the following theorem.

Theorem. *Any algebraic sum is equal to the sum of those of its elements which are preceded by the plus sign minus the sum of those which ar preceded by the minus sign.*

261. Cancellation. Suppose we hav an algebraic sum two of whose elements ar equal, but one preceded by the plus sign and the other by the minus sign. Then these integers with their signs may be removed from the given expression.

For, suppose the equal elements with their signs ar $+\alpha$ and $-\beta$. The order of the elements of our sum may be changed so as to bring these integers to the end of the expression. Represent all of the algebraic sum thus obtaind except the $+\alpha$ and $-\beta$ by μ. The algebraic sum is then reduced to the form $\mu+\alpha-\beta$. This may now be written $\mu+(\alpha-\beta)$, or $\mu+0$ (§ 233), which equals μ.

Thus in the algebraic sum considerd in § 260, since $\beta=\beta$, β and $-\beta$ may be omitted; similarly $+\delta$ and $-\delta$ may be omitted. This may be indicated by drawing a line thru these letters, or **cancelling** [1] them, thus:

$$\not{\beta}+\gamma+\not{\delta}+\theta-\alpha-\varepsilon-\zeta-\eta-\not{\delta}-\not{\beta}.$$

The given expression is, then, equal to

$$\gamma+\theta-\alpha-\varepsilon-\zeta-\eta.$$

We may state the result proved above as follows.

Theorem. *If in any algebraic sum two elements ar equal, but one is preceded by the plus sign and the other by the minus sign, these elements with their signs may be removed from the given expression.*

---

(1) See Fisher and Schwatt's "Text-Book of Algebra," 1898, Part I, p. 25.

TRANSPOSITION.

262. Theorem. *If* $\alpha = \beta$, *then* $\alpha - \beta = 0$.

Since $\alpha = \beta$, $\alpha - \beta = \beta - \beta$. § 237.

Or $\alpha - \beta = 0$. § 228.

263. Theorem. *If* $\alpha = -\beta$, *then* $\alpha + \beta = 0$.

For $\alpha - (-\beta) = \alpha + \beta$. § 225.

264. Theorem. *If* $\alpha = \beta + \gamma$, *then* $\alpha - \gamma = \beta$.

Since $\alpha = \beta + \gamma$, $\alpha - \gamma = \beta + \gamma - \gamma$. § 237.

Or $\alpha - \gamma = \beta$. § 261.

265. Theorem. *If* $\alpha = \beta - \gamma$, *then* $\alpha + \gamma = \beta$.

266. Theorem. *Any integer in either member of an equality may be transposed* (1) *to the other, if its sign is changed, the member from which the integer is removed becoming zero, if the integer transposed was the entire member.*

This follows immediately from §§ 262–265.

267. Theorem. *The signs of all the integers in an equality may be changed.*

This is proved by transposing all the integers, by § 266, and then reversing the equality.

Thus, if $\alpha + \beta = -\gamma + \delta$,

$$\gamma - \delta = -\alpha - \beta.$$

Hence $-\alpha - \beta = \gamma - \delta$.

---

(1) See Beman and Smith, "Elements of Algebra," 1900, p. 148.

268. Theorem. *If* $\alpha > \beta$, *then* $\alpha - \beta > 0$.

This is proved in the same way as § 262.

269. Theorem. *If* $\alpha > -\beta$, *then* $\alpha + \beta > 0$.

270. Theorem. *If* $\alpha > \beta + \gamma$, *then* $\alpha - \gamma > \beta$.

271. Theorem. *If* $\alpha > \beta - \gamma$, *then* $\alpha + \gamma > \beta$.

272. Theorem. *If* $\alpha < \beta$, *then* $\alpha - \beta < 0$.

273. Theorem. *If* $\alpha < -\beta$, *then* $\alpha + \beta < 0$.

274. Theorem. *If* $\alpha < \beta + \gamma$, *then* $\alpha - \gamma < \beta$.

275. Theorem. *If* $\alpha < \beta - \gamma$, *then* $\alpha + \gamma < \beta$.

276. Theorem. *Any integer in either member of an inequality may be transposed to the other, if the sign of the integer is changed, the member from which the integer is removed becoming zero, if the integer transposed was the entire member.*

277. Theorem. *The signs of all the integers in an inequality may be changed, if the sign of inequality is changed, from* $>$ *to* $<$ *or from* $<$ *to* $>$.

Thus, if $\alpha - \beta > -\gamma - \delta$,

$$\gamma + \delta > -\alpha + \beta.$$

Hence $-\alpha + \beta < \gamma + \delta$.

278. Theorem. *Equal integers, one in each member of an equality or inequality and preceded by the same sign, may be canceld.*

This follows from §§ 266, 276, 261.

279. Integers orderd. We will now arrange our system of integers in a series so that each shall be one less than the succeeding one. We hav seen that in the natural series of primary numbers each number is one less than the immediately succeeding number (§ 133). Starting with any given primary number we could therfor, by successivly subtracting ones from it, obtain all the primary numbers below it, down to one. But we could not continue beyond this point as ther is no primary number smaller than one.

With integers, however, we can continue this process as far as we please.

Thus $$1 - 1 = 0,\ 0 - 1 = \bar{1},$$
$$\bar{1} - 1 = -(1 + 1) = \bar{2},\ \bar{2} - 1 = -(2 + 1) = \bar{3},\ \cdots$$

Thus we hav the series

$$\cdots, \bar{9}, \bar{8}, \bar{7}, \bar{6}, \bar{5}, \bar{4}, \bar{3}, \bar{2}, \bar{1}, 0, 1, 2, 3, 4, 5, 6, 7, 8, 9, \cdots$$

If $\bar{a}$ is any negativ integer in this series, the one preceding it is $\bar{a} - 1$ or $-(a + 1)$. Hence the opposit of every primary number will be found somewher in the series (§ 65). And since every negativ integer is the opposit of some primary number, every negativ integer will be found in the series. We also hav zero and every positiv integer in the series. Therfor the series includes all the integers (1).

280. Definition. The series given above may be calld the **natural series of integers.** The number 0 is calld the **center** of the series. The other integers may

(1) Negativ integers and zero seem to hav been freely used as numbers first by Descartes in the seventeenth century. See his "La Géométrie," 1637. The first traces of such use ar, however, found in the work on astronomy of Bhâskara, an Indian mathematician of the twelfth century.

be obtaind by adding to and subtracting from zero in succession the natural series of primary numbers. §§ 209, 230.

281. Theorem. *In the natural series of integers each integer is one greater than the preceding integer.*

For, if $\alpha = \beta - 1$, $\beta - \alpha = 1$.

282. Theorem. *Ther is no greatest integer and ther is no smallest integer.*

The first part of the theorem follows from the fact that every positiv integer is a primary number and is greater than zero and greater than every negativ integer.

The second part follows from the fact that every negativ integer is less than zero and less than every positiv integer, and that $-(a+1) < -a$. § 235.

This theorem is equivalent to the statement that *the series of integers has no beginning and no end; it is limitless, or infinit, in both directions.*

283. Theorem. *If α is any integer in the natural series, the integers that immediately precede α ar in order $\alpha - 1, \alpha - 2, \alpha - 3, \cdots, \alpha - n, \alpha - (n+1), \cdots$, and those that immediately follow α ar in order $\alpha + 1$, $\alpha + 2$, $\alpha + 3, \cdots$, $\alpha + n$, $\alpha + (n+1), \cdots$, wher $1, 2, 3, \cdots$, n, $n+1$, $\cdots$ is the natural series of numbers.*

Let the integers that immediately precede α in order be $\beta, \gamma, \delta, \cdots, \kappa, \lambda, \cdots$

Then $\beta = \alpha - 1$, § 279.

$$\gamma = \beta - 1 = (\alpha - 1) - 1 = \alpha - (1 + 1) = \alpha - 2,$$ § 260.

$$\delta = \gamma - 1 = (\alpha - 2) - 1 = \alpha - (2 + 1) = \alpha - 3,$$

. . . .

Moreover, if $\kappa = \alpha - n$,

$$\lambda = \kappa - 1 = (\alpha - n) - 1 = \alpha - (n + 1).$$

Hence the integers $\beta, \gamma, \delta, \cdots, \kappa, \lambda, \cdots$ ar equal respectivly to the integers $\alpha - 1,\ \alpha - 2,\ \alpha - 3,\ \cdots,$ $\alpha - n,\ \alpha - (n + 1), \cdots$

In this series the numbers subtracted from α ar each, after the first, obtaind from the preceding by adding 1 to it. These numbers ar therfor the natural series of numbers.

The last part of the theorem is proved in the same way as § 69.

Thus the natural series of integers may be written with α as a center as follows:

$$\cdots, \alpha - (n + 1),\ \alpha - n, \cdots, \alpha - 3,\ \alpha - 2,\ \alpha - 1,\ \alpha,$$
$$\alpha + 1,\ \alpha + 2,\ \alpha + 3, \cdots, \alpha + n,\ \alpha + (n + 1), \cdots$$

284. Theorem. *If $\alpha < \beta < \gamma$, β is one of the numbers in the natural series of integers between α and γ.*

Let $\beta - \alpha = \delta$ and $\gamma - \beta = \varepsilon$.

Then $\beta = \alpha + \delta$ and $\gamma = \beta + \varepsilon$, wher δ and ε ar positiv. §§ 266, 234.

The proof is now the same as that of § 139.

285. Definition. When $\alpha < \beta < \gamma$, β is said to **lie between** α and γ in magnitude.

286. Theorem. *If α and β ar any two consecutiv integers in the natural series of integers, ther exists no integer between α and β.*

Suppose $\alpha < \gamma < \beta$.

Then $\beta - \alpha = \beta - \gamma + \gamma - \alpha$ § 261.

$= (\beta - \gamma) + (\gamma - \alpha),$ § 257.

wher $\beta - \gamma$ and $\gamma - \alpha$ ar positiv. § 234.

Therfor $\beta - \alpha > \beta - \gamma \geqq 1$.

The proof is now the same as that of § 141.

287. Definition. In the natural series of integers the alternate integers commencing with zero and going either way ar **even**; the others ar **odd**. This agrees with the definition of § 68 for primary numbers.

Thus

$\cdots, \bar{8}, \bar{6}, \bar{4}, \bar{2}, 0, 2, 4, 6, 8, \cdots$ ar even;

$\cdots, \bar{9}, \bar{7}, \bar{5}, \bar{3}, \bar{1}, 1, 3, 5, 7, 9, \cdots$ ar odd.

288. Theorem. *If α is even, $\bar{\alpha}$ is even; if α is odd $\bar{\alpha}$ is odd.*

289. Theorem. *The sum of two even integers, or of two odd integers, is even; the sum of two integers, one of which is even and the other odd, is odd; and conversely, if the sum of two integers is even, they ar either both even or both odd; if the sum of two integers is odd, one must be even and the other odd.*

For the direct theorem, suppose first that α is an even integer.

Now, if β is zero, $\alpha + \beta = \alpha$ and $\alpha + \beta$ will be even. § 182.

If β is positiv and equal to b, consider the portion of the natural series of integers

$$\alpha + 1, \alpha + 2, \alpha + 3, \cdots, \alpha + b.$$

Since α is even, the first of these numbers is odd.

Hence the second is even, the third odd, etc.

If we make these numbers correspond to the numbers

$$1, 2, 3, \cdots, b,$$

we see that the numbers of the first series which correspond to odd numbers in the second ar odd and those that correspond to even numbers in the second ar even.

Therfor $\alpha + \beta$ is odd or even, according as β is odd or even.

If β is negativ and equal to $-b$, we consider the series of integers

$$\alpha - 1, \alpha - 2, \alpha - 3, \cdots, \alpha - b$$

and see in a similar manner that $\alpha + \beta$, or $\alpha - b$, is odd or even according as β is odd or even.

Therfor, in all cases, if α is even, $\alpha + \beta$ is odd or even according as β is odd or even.

The case when α is odd is proved similarly.

Each part of the converse theorem may be proved by the method of exhaustion.

For the first part, when the sum is even, ther ar only three possible cases.

Either one integer is even and the other odd, or both ar even, or both ar odd.

If one were even and the other odd, the sum would be odd, by the direct theorem.

But, by the hypothesis, the sum is not odd.

Hence we cannot hav one even and the other odd.

Therfor they ar either both even or both odd.

The syllogism for the last step of this proof is an amplified form of syllogism V of § 131. It may be written schematically as follows:

Major Premis. Either A is B, or C is D, or E is F.

Minor Premis. A is not B.

Conclusion. Therfor either C is D, or E is F.

For the second part of the converse theorem, the same three cases ar to be considerd.

We may show by the indirect method that the last two ar impossible.

Therfor the first case must be true.

The syllogism used here may be written, changing the order of the cases:

Major Premis. Either A is B, or C is D, or E is F.

Minor Premis. A is not B and C is not D.

Conclusion. Therfor E is F.

290. Theorem. *The difference of two even integers, or of two odd integers, is even; the difference of two integers, one of which is even and the other odd, is odd; and conversely, if the difference of two integers is even, they ar either both even or both odd; if the difference is odd, one must be even and the other odd.*

This may be proved indirectly using § 265.

291. Theorem. *The sum of any number of even integers is even; the sum of an even number of odd integers is even; the sum of an odd number of odd integers is odd.*

The first part of the theorem follows from § 289 by the method used to prove § 254.

For the second part, if we add the odd integers together in pairs, the original sum will be the same as the sum of a number of even integers (§ 289). Hence this part of the theorem follows from the first part.

To prove the last part, the sum of all the integers except one will be the sum of an even number of odd integers, and hence even. Hence this sum plus the remaining odd integer will be odd. § 289.

292. Theorem. *An algebraic sum of any number of even integers is even; an algebraic sum of an even number of odd integers is even; an algebraic sum of an odd number of odd integers is odd.*

This is proved by changing all the minus signs befor integers in the algebraic sum to plus, changing also the signs over these integers, and then applying §§ 288, 291.

293. Theorem. *If α and β ar two successiv even integers or two successiv odd integers in the natural series of integers, $\beta = \alpha + 2$; and conversely, if $\beta = \alpha + 2$ and α is even, β is even; if α is odd, β is odd.*

For the first part of the theorem, $\beta = \alpha + 1 + 1$. Hence the theorem follows.

The converse theorem is proved by reversing the steps.

294. Theorem. *Every even integer can be obtaind from every other even integer and every odd integer can be obtaind from every other odd integer by adding to,* *or subtracting from, it a certain number of twos; and conversely, every integer that can be obtaind from another integer by adding to, or subtracting from, it a number of twos is even or odd according as that integer is even or odd.*

For, if κ and α ar two even integers and κ comes somewher after α in the natural series of integers, the even integer next after α is $\alpha + 2$ (§ 293), the next one $\alpha + 2 + 2$, the next $\alpha + 2 + 2 + 2$, and so on. And κ is one of this series of numbers.

If κ comes somewher befor α, the even integer next befor α is $\alpha - 2$, the one befor it $\alpha - 2 - 2$, the next $\alpha - 2 - 2 - 2$, and so on.

Similarly if κ and α ar odd integers.

The converse theorem is proved by applying § 293.

295. Definition. The system of integers is calld a **closed system** with reference to the operations of addition and subtraction, because each of these operations can be performd on any two integers and the result of the operation is always a number of the system.

CHAPTER IV.

MULTIPLICATION.

296. Definition. Product. By the generalized associativ law for addition of integers the sum of a group of n a's, wher a is any integer and n a primary number, is the same, however we group the a's in parentheses in finding their sum. This sum, then, is univalent and is determind by a and n (¹). We will therfor call it the **product** (²) of a and n.

This definition implies that $n > 1$. For to hav a sum we must hav at least two elements. But just as we speak of a group consisting of only one object, so we will speak of adding a group of n a's, even when $n = 1$, the result of the addition being a itself.

Thus the product of a and n may in all cases be defined as the sum of a group of n a's.

The product of a and n may be indicated by either of the symbols $a \times n$, $a \cdot n$ (³), or simply an (⁴),

---

(¹) See Stolz und Gmeiner, p. 21.

(²) See Schubert, Encyklopädie, p. 14.

(³) See Euler, p. 9.

(⁴) The sign $\times$ to indicate a product was apparently first used by Wm. Oughtred in 1631. See his "Clavis Mathematica Denuo Limata," 1648, p. 10. The sign · was used by T. Harriot in 1631 to denote a product. See his "Artis Analyticæ Praxis," 1631. Stifel in 1544 in his "Arithmetica Integra" indicated the same by juxtaposition.

which ar read "a times n," or simply "an." Thus an stands for $a + a + a + \cdots$ (n times).

The integer a is calld the **base** of the product, the primary number n the **coefficient.**

The product an is also calld a **multiple**, an **n-multiple,** of a; $a \times 1$ is the **unit multiple** of a.

297. Definition. Operation of Multiplication. The product an is obtaind from a and n by an operation, the operation of adding n a's. This operation is calld **multiplication** and a is said to be **multiplied** by n. In this operation a is the operand, passiv element, or **multiplicand**, and n the operator, activ element, or **multiplier** (1).

298. Theorem. *The product an is positiv if a is positiv, zero if a is zero, negativ if a is negativ.*

This follows immediately from § 254.

299. Theorem. $a \times 1 = a$.

For $a \times 1$ stands for a. Hence $a \times 1 = a$. § 170.

300. Theorem. $1 \times 1 = 1$.

301. Theorem. $a \times n = a + a + a + \cdots$ (*n times*).

For $a \times n$ stands for $a + a + a + \cdots$ (n times).

Hence $a \times n = a + a + a + \cdots$ (n times). § 170.

302. Theorem. $1 \times n = n$.

This follows immediately from §§ 301, 63.

303. Theorem. $0 \times n = 0$.

This follows immediately from §§ 301, 255.

304. Theorem. $(-a) n = -(an)$.

---

(1) See Schubert, Encyklopädie, p. 14.

For $\bar{a} + \bar{a} + a + \cdots n$ times $=$

$$-(a + a + a + \cdots n \text{ times}). \qquad § 256.$$

That is, $\quad (-a)n = -(an). \qquad § 175.$

305. Theorem. *If either a or n is even, an is even; if both a and n ar odd, an is odd.*

This follows immediately from §§ 301, 291.

306. Theorem. *If $a = \beta$, $an = \beta n$.*

This is proved by adding together n equalities of the form $a = \beta$. §§ 252, 301, 175.

307. Theorem. *If $a > \beta$, $an > \beta n$.*

This is proved by adding together n inequalities of the form $a > \beta$. § 253.

308. Theorem. *If $m = n$, $am = an$.*

For am represents the sum of m a's and an the sum of n a's.

Since $m = n$, ther ar as many a's in the one group as in the other. § 5.

Hence the expressions am and an ar different symbols for the same sum.

Therfor $am = an$. § 170.

309. As a product is an integer, we may multiply it by a primary number or we may combine it with other products or integers by plus or minus signs, using parentheses to indicate the order in which the various operations ar performd. Thus we may hav expressions like $(am) + (\beta n)$ and $a((l + m)n)$.

All such expressions ar univalent. For every sum, difference, and product is univalent. §§ 203, 220, 296.

Generally, in complex expressions involving multiplication with addition or subtraction, or both addition and subtraction, parentheses ar omitted from around products[1]. They should never be omitted from around sums or differences, when such omission would cause misunderstanding. Thus $al + mn$ is understood to mean $(al) + (mn)$ and not $a[(l + m)n]$.

310. Theorem. $am + an = a(m + n)$[2].

For $am + an = (a + a + a + \cdots m$ times)
$+ (a + a + a + \cdots n$ times) §§ 301, 214.

$= a + a + a + \cdots (m + n)$ times §§ 250, 37.

$= a\,(m + n)$. § 301.

Therfor

$am + an = a(m + n)$. § 175.

This theorem is the **Left-handed Distributiv**[3] **Law for Multiplication and Addition.**

311. Theorem. $\alpha n + \beta n = (\alpha + \beta)n$.

For $\alpha n + \beta n = (\alpha + \alpha + \alpha + \cdots n$ times)
$+ (\beta + \beta + \beta + \cdots n$ times)

$= (\alpha + \beta) + (\alpha + \beta) + (\alpha + \beta) + \cdots n$ times

$= (\alpha + \beta)n$.

Therfor $\alpha n + \beta n = (\alpha + \beta)n$.

---

(1) See foot-note p. 85.

(2) See Schubert, Encyklopädie, p. 14; Tannery, p. 57.

(3) The term "distributiv," in this sense, was first used by Servois, Gergonne's *Annales*, Vol. V, 1814, p. 98.

This theorem is the **Right-handed Distributiv Law for Multiplication and Addition.**

312. Theorem. *If* $m > n$, $am - an = a(m - n)$; *and, if* a *is positiv,* $am > an$; *if* a *is negativ,* $am < an$.

Since $m > n$, $m = x + n$ and $m - n = x$.

Hence $am = a(x + n)$. § 308.

Or $am = ax + an$ and $am - an = ax$. §§ 310, 266.

Hence $am - an = a(m - n)$. §§ 308, 173.

Moreover, if a is positiv, ax is positiv. § 298.

Hence $am - an$ is positiv and $am > an$. §§ 176, 234.

Similarly, if a is negativ, $am < an$.

313. Definition. According to the definition for a product given in § 296, the multiplicand may be any integer, but the multiplier must be a primary number. We ar now ready to extend the definition to the case when the multiplier is any integer.

By § 312, when $b_1 > b_2$, the product $a(b_1 - b_2)$ is $ab_1 - ab_2$; that is, when $b_1 > b_2$, $a(b_1, b_2)$ represents $ab_1 - ab_2$ (§ 240). Moreover the symbol $ab_1 - ab_2$ is univalent in all cases. §§ 296, 220.

If, therfor, we make $ab_1 - ab_2$ the product of a and (b_1, b_2) in all cases, our definitions will not be contradictory.

Formally, then, we define: $a(b_1, b_2)$ represents $ab_1 - ab_2$ (1).

(1) See Schubert, Encyklopädie, p. 15.

314. Operation of Multiplication of Integers.

The product $\alpha\beta$ of two integers is obtaind by an operation. For ab_1 and ab_2 ar each obtaind by an operation (§ 297) and then ab_2 is subtracted from ab_1.

This operation is calld **multiplication of integers**; α is the **multiplicand** and β the **multiplier.**

315. Theorem. $\alpha\beta = ab_1 - ab_2$.

For $\alpha\beta$ represents $ab_1 - ab_2$ and hence

$$\alpha\beta = ab_1 - ab_2. \qquad \text{§ 170.}$$

316. Theorem. $1 \times \alpha = \alpha$.

For $1 \times \alpha = 1 \times a_1 - 1 \times a_2$ § 315.

$= a_1 - a_2$ §§ 302, 239.

$= \alpha$. § 240.

Therfor $1 \times \alpha = \alpha$. § 175.

317. Theorem. $0 \times \alpha = 0$.

For $0 \times \alpha = 0 \times a_1 - 0 \times a_2$

$= 0 - 0$

$= 0$.

318. Theorem. $\alpha \times 0 = 0$.

For $\alpha \times 0 = ac_1 - ac_2$, wher $c_1 = c_2$. § 181.

Since $c_1 = c_2$, $ac_1 = ac_2$. § 308.

Hence $ac_1 - ac_2 = 0$. § 262.

Therfor $\alpha \times 0 = 0$. § 173.

319. Theorem. $0 \times 0 = 0$.

320. Theorem. *The product of two positiv, or of two negativ integers, is positiv; the product of two integers, one of which is zero, is zero; the product of two integers, of which one is positiv and the other negativ, is negativ; and conversely, if the product of two integers is positiv, they ar either both positiv or both negativ; if the product is zero, one of them must be zero; if the product is negativ, one must be positiv and the other negativ.*

Suppose, first, that α and β ar both positiv.

Then $b_1 > b_2$. § 159.

Hence $ab_1 > ab_2$. § 312.

Therfor $ab_1 - ab_2$ is positiv. § 234.

That is, $\alpha\beta$ is positiv. §§ 315, 176.

The other cases of the direct theorem ar proved in a similar way.

The converse theorem is proved by the method of exhaustion.

321. Theorem. $(-\alpha)\beta = -(\alpha\beta)$.

For $(-\alpha)\beta = (-a)b_1 - (-a)b_2$ § 315.

$= -(ab_1) - (-(ab_2))$ §§ 304, 239.

$= -(ab_1 - ab_2)$ § 232.

$= -(\alpha\beta)$. §§ 315, 177.

Therfor $(-\alpha)\beta = -(\alpha\beta)$. § 175.

322. Since the expressions $(-\alpha)\beta$ and $-(\alpha\beta)$ ar equal, the parentheses may be omitted and both be written $-\alpha\beta$.

323. Theorem. $\alpha(-\beta) = -(\alpha\beta)$.

For $\alpha(-\beta) = \alpha(b_2, b_1) = ab_2 - ab_1$

$$= -(ab_1 - ab_2).$$

324. Theorem. $(-\alpha)(-\beta) = \alpha\beta$.

For $(-\alpha)(-\beta) = -(\alpha(-\beta))$ § 321.

$= -(-(\alpha\beta))$ §§ 323, 177.

$= \alpha\beta$. § 184

325. Theorem. $(-1)\beta = -\beta$.

326. From the theorems of §§ 170, 321, 323, 324 we derive the following equalities:

$$(+\alpha)(+\beta) = +\alpha\beta,$$

$$(-\alpha)(+\beta) = -\alpha\beta,$$

$$(+\alpha)(-\beta) = -\alpha\beta,$$

$$(-\alpha)(-\beta) = +\alpha\beta.$$

These equalities may be included in the following statement.

Theorem. *If a product consists of two factors each of which is an integer preceded by a plus or a minus sign, the product is equal to the product of the given integers preceded by a plus sign, if the signs of the two integers ar alike, but by a minus sign, if those signs ar unlike.*

This theorem is the **Rule of Signs for Multiplication.**

It is usually briefly stated: *In multiplication like signs giv +, unlike signs —.*

In particular we hav $(-a)b = a(-b) = -(ab)$ and $(-a)(-b) = ab$.

327. Theorem. *If γ is zero, $\alpha\gamma = \beta\gamma$.*

For then $\alpha\gamma$ and $\beta\gamma$ ar both zero. § 320.

328. Theorem. *If γ is zero, $\gamma\alpha = \gamma\beta$.*

329. Theorem. *If $\alpha = \beta$, $\alpha\gamma = \beta\gamma$.*

Since $\alpha = \beta$, $\alpha c_1 = \beta c_1$ § 306.

and $\alpha c_2 = \beta c_2$.

Hence $\alpha c_1 - \alpha c_2 = \beta c_1 - \beta c_2$. § 239.

That is, $\alpha\gamma = \beta\gamma$. § 315.

330. Theorem. *If $\alpha = \beta$, $\gamma\alpha = \gamma\beta$.*

Since $\alpha = \beta$, $a_1 + b_2 = a_2 + b_1$. § 168.

Hence $\gamma(a_1 + b_2) = \gamma(a_2 + b_1)$. § 308.

Or $\gamma a_1 + \gamma b_2 = \gamma a_2 + \gamma b_1$. § 310.

Therfor $\gamma a_1 - \gamma a_2 = \gamma b_1 - \gamma b_2$. § 266.

That is, $\gamma\alpha = \gamma\beta$.

331. Theorem. *If $\alpha = \beta$ and $\gamma = \delta$, then $\alpha\gamma = \beta\delta$.*

This is proved in the same way as § 214.

This theorem is usually briefly stated: *Equals multiplied by equals giv equals.*

332. Theorem. *If $\alpha > \beta$ and γ is positiv, $\alpha\gamma > \beta\gamma$.*

Since γ is positiv, it is a primary number c.

Since $\alpha > \beta$, $\alpha c > \beta c$. § 307.

But, since $\gamma = c$, $\alpha\gamma = \alpha c$

and $\beta c = \beta\gamma$. § 330.

Therfor $\alpha\gamma > \beta\gamma$. § 197.

333. Theorem. *If* $\alpha > 1$, $\alpha\alpha > \alpha$.

This follows immediately from § 332 by putting 1 instead of β and α instead of γ.

334. Theorem. *If* ν *is any positiv integer, ther is a positiv integer* α *such that* $\alpha\alpha > \nu$.

If $\nu > 1$, this theorem follows immediately from the preceding.

If $\nu = 1$, we hav $2 \times 2 > 2$ by § 333.

Hence, since $2 > 1$, we hav $2 \times 2 > \nu$.

335. Theorem. *If* $\alpha > \beta$ *and* γ *is positiv,* $\gamma\alpha > \gamma\beta$.

Since $\alpha > \beta$, $a_1 + b_2 > a_2 + b_1$. § 186.

Hence $\gamma(a_1 + b_2) > \gamma(a_2 + b_1)$. § 312.

Or $\gamma a_1 + \gamma b_2 > \gamma a_2 + \gamma b_1$. § 310.

Therfor $\gamma a_1 - \gamma a_2 > \gamma b_1 - \gamma b_2$. § 276.

That is, $\gamma\alpha > \gamma\beta$. § 315.

336. Theorem. *If* $\alpha > \beta$ *and* γ *is negativ,* $\alpha\gamma < \beta\gamma$.

Since γ is negativ, $-\gamma$ is positiv. § 164.

Hence, since $\alpha > \beta$,

$\alpha(-\gamma) > \beta(-\gamma)$. § 332.

Or $-(\alpha\gamma) > -(\beta\gamma)$. § 323.

Therfor $\alpha\gamma < \beta\gamma$. § 189.

337. Theorem. *If* $\alpha > \beta$ *and* γ *is negativ,* $\gamma\alpha < \gamma\beta$.

338. From the theorems of §§ 332, 335–337 we deduce the following:

Theorem. *If* $\alpha \neq \beta$ *and* $\gamma \neq 0$, $\alpha\gamma \neq \beta\gamma$.

For then either $\alpha > \beta$ or $\alpha < \beta$, and γ is either positiv or negativ.

Hence either $\alpha\gamma > \beta\gamma$ or $\alpha\gamma < \beta\gamma$.

339. Theorem. Hence, *if* $\alpha\gamma = \beta\gamma$, *either* $\alpha = \beta$ *or* $\gamma = 0$.

340. Theorem. *If* $\alpha \neq \beta$ *and* $\gamma \neq 0$, $\gamma\alpha \neq \gamma\beta$.

341. Theorem. *If* $\gamma\alpha = \gamma\beta$, *either* $\gamma = 0$ *or* $\alpha = \beta$.

342. Theorem. *If* $\alpha \neq 0$ *and* $\beta \neq 0$, $\alpha\beta \neq 0$.

This is proved from § 338 by replacing the β and γ of that theorem by 0 and β respectivly.

343. From this we may deduce the following important theorem.

Theorem. *If* $\alpha\beta = 0$, *then either* $\alpha = 0$ *or* $\beta = 0$.

344. Theorem. From the last theorem and §§ 317, 318 we conclude that *the product of two integers equals zero when, and only when, one of these integers is equal to zero.*

345. Theorem. *If* $\alpha\gamma > \beta\gamma$ *and* γ *is positiv*, $\alpha > \beta$.

For either $\alpha = \beta$, or $\alpha < \beta$, or $\alpha > \beta$. § 187.

If α were equal to β, $\alpha\gamma$ would equal $\beta\gamma$. § 329.

But $\alpha\gamma$ is given greater than $\beta\gamma$ and hence cannot equal $\beta\gamma$. § 187.

Therfor α cannot equal β.

Similarly it may be shown that α cannot be less than β.

Therfor $\alpha > \beta$.

346. Theorem. *If* $\alpha\gamma > \beta\gamma$ *and* γ *is negativ,* $\alpha < \beta$.

347. Theorem. *If* $\gamma\alpha > \gamma\beta$ *and* γ *is positiv,* $\alpha > \beta$.

348. Theorem. *If* $\gamma\alpha > \gamma\beta$ *and* γ *is negativ,* $\alpha < \beta$.

349. Theorem. *If* $\alpha > \beta$ *and* $\gamma = \delta$; *then, if* γ *is positiv,* $\alpha\gamma > \beta\delta$; *if* γ *is negativ,* $\alpha\gamma < \beta\delta$.

This is proved like § 217.

350. Theorem. *If* $\alpha > \beta$ *and* $\gamma > \delta$; *then, if* α *and* δ, *or* β *and* γ, *ar both positiv,* $\alpha\gamma > \beta\delta$; *if* α *and* δ, *or* β *and* γ, *ar both negativ,* $\alpha\gamma < \beta\delta$.

If α and δ ar both positiv, since $\gamma > \delta$,

$$\alpha\gamma > \alpha\delta. \qquad \text{§ 335.}$$

Since $\alpha > \beta$, $$\alpha\delta > \beta\delta. \qquad \text{§ 332.}$$

Therfor $$\alpha\gamma > \beta\delta. \qquad \text{§ 191.}$$

Similarly the other parts of the theorem ar proved.

351. Theorem. *If* $\alpha \neq \beta$ *and* $\gamma = \delta$, *then* $\alpha\gamma \neq \beta\delta$, *provided* $\gamma \neq 0$.

For, if $\alpha > \beta$, $\alpha\gamma \gtrless \beta\delta$ according as γ is positiv or negativ.

Similarly if $\alpha < \beta$.

Therfor in both cases $\alpha\gamma \neq \beta\delta$.

352. Theorem. *If* $\alpha\gamma = \beta\delta$ *and* $\gamma = \delta$, *then* $\alpha = \beta$, *provided* $\gamma \neq 0$.

353. Theorem. *If* $\alpha\gamma = \beta\delta$ *and* $\alpha > \beta$, *then, if* α *and* δ, *or* β *and* γ, *ar both positiv, or both negativ,* $\gamma < \delta$.

This may be proved by the method of exhaustion.

354. Theorem. *If* $\alpha\gamma > \beta\delta$ *and* $\alpha = \beta$; *then, if* α *is positiv,* $\gamma > \delta$; *if* α *is negativ,* $\gamma < \delta$.

355. Theorem. *If* $\alpha\gamma > \beta\delta$ *and* $\alpha < \beta$*; then, if* α *and* δ*, or* β *and* γ*, ar both positiv,* $\gamma > \delta$.

356. Theorem. *If* $\alpha\gamma > \beta\delta$ *and* $\alpha > \beta$*; then, if* α *and* δ*, or* β *and* γ*, ar both negativ,* $\gamma < \delta$.

357. Theorem. *If* $\alpha\alpha > \beta\gamma$ *and* α *is positiv, either* $\alpha > \beta$ *or* $\alpha > \gamma$*; if* $\alpha\alpha > \beta\gamma$ *and* α *is negativ, either* $\alpha < \beta$ *or* $\alpha < \gamma$.

For the first part of the theorem, either $\alpha > \beta$, or $\alpha = \beta$, or $\alpha < \beta$.

In the first case the theorem is true.

If $\alpha = \beta$, then $\alpha > \gamma$. § 354.

If $\alpha < \beta$, then β must also be positiv. § 201.

Hence $\alpha > \gamma$. § 355.

The second part of the theorem is proved similarly.

358. Theorem. *If* $\alpha\beta > \gamma\gamma$ *and* α *is positiv, either* $\alpha > \gamma$ *or* $\beta > \gamma$*; if* $\alpha\beta > \gamma\gamma$ *and* α *is negativ, either* $\alpha < \gamma$ *or* $\beta < \gamma$.

359. Theorem. *If* $\alpha\alpha > \beta\beta$*, then, if* α *is positiv,* $\alpha > \beta$*; if* α *is negativ,* $\alpha < \beta$.

This follows immediately from § 357 or § 358.

360. Theorem. *If either of two integers is even, their product is even; if both ar odd, their product is odd; and conversely, if the product of two integers is even, one of them must be even; if the product is odd, both must be odd.*

Let α and β be the two integers, $\alpha\beta$ their product.

The direct theorem has already been proved in § 305 when β is positiv.

Suppose β is zero. Then $\alpha\beta$ is zero and the theorem is true. §§ 320, 287.

Suppose β is negativ and equal to $-n$, wher n is even or odd according as β is even or odd. § 288.

Then $\alpha\beta = -(\alpha n)$. §§ 330, 323.

But αn is even, if either α or n is even, and odd, if both α and n ar odd. § 305.

Therfor $\alpha\beta$ is even, if either α or β is even, and odd, if both α and β ar odd. §§ 288, 182.

The converse theorem may be proved by the method of exhaustion.

361. As a product is **an integer,** we may multiply it by another integer or we may combine it with other products or integers by plus or minus signs, using parentheses to indicate the order in which the various operations ar performd. Thus we may hav expressions like $(\alpha\beta) + (\gamma\delta)$ and $(\alpha(\beta + \gamma))\delta$.

362. Theorem. *If for one or more elements of a complex expression containing integers connected by signs of multiplication, with or without signs of addition or subtraction, or both, equal integers ar substituted, the complex expression is unchanged.*

363. Theorem. $\alpha\beta + \alpha\gamma = \alpha(\beta + \gamma)$.

For $\alpha\beta + \alpha\gamma = (ab_1 - ab_2) + (ac_1 - ac_2)$. §§ 315, 214.

$= (ab_1 + ac_1) - (ab_2 + ac_2)$ § 259.

$= a(b_1 + c_1) - a(b_2 + c_2)$ §§ 310, 239.

$= a(b_1 + c_1, b_2 + c_2)$ § 315.

$= \alpha(\beta + \gamma)$. § 170.

364. Theorem. $\alpha\gamma + \beta\gamma = (\alpha + \beta)\gamma$.

365. Theorem. $(\alpha + \beta)(\gamma + \delta) = \alpha\gamma + \beta\gamma + \alpha\delta + \beta\delta$.

Treating $\alpha + \beta$ first as a single symbol, we hav

$(\alpha + \beta)(\gamma + \delta) = (\alpha + \beta)\gamma + (\alpha + \beta)\delta$ § 363.

$= \alpha\gamma + \beta\gamma + \alpha\delta + \beta\delta$. §§ 364, 214.

366. Theorem. $\alpha\beta - \alpha\gamma = \alpha(\beta - \gamma)$.

For $\alpha\beta - \alpha\gamma = \alpha\beta + \alpha\bar{\gamma}$ §§ 223, 323, 213.

$= \alpha(\beta + \bar{\gamma})$ § 363.

$= \alpha(\beta - \gamma)$ §§ 223, 330.

367. Theorem. $\alpha\gamma - \beta\gamma = (\alpha - \beta)\gamma$.

368. Theorem. $(\alpha - \beta)(\gamma - \delta) = \alpha\gamma - \beta\gamma - \alpha\delta + \beta\delta$.

369. Theorem. $\nu(\alpha + \beta + \gamma + \cdots + \kappa + \lambda + \mu)$
$= \nu\alpha + \nu\beta + \nu\gamma + \cdots + \nu\kappa + \nu\lambda + \nu\mu$.

To prove this we write $\alpha + \beta + \gamma + \cdots + \kappa + \lambda + \mu$ as a standard sum and apply § 363.

Thus $\nu(((((\alpha + \beta) + \gamma) + \cdots + \kappa) + \lambda) + \mu)$

$= \nu((((\alpha + \beta) + \gamma) + \cdots + \kappa) + \lambda) + \nu\mu$

$= \nu(((\alpha + \beta) + \gamma) + \cdots + \kappa) + \nu\lambda + \nu\mu$,

and so on.

This theorem is the **Generalized Left-handed Distributiv Law for Multiplication and Addition.**

370. Theorem. $(\alpha + \beta + \gamma + \cdots + \kappa + \lambda + \mu)\nu$
$= \alpha\nu + \beta\nu + \gamma\nu + \cdots + \kappa\nu + \lambda\nu + \mu\nu$.

Proved from § 364 like § 369.

This theorem is the **Generalized Right-handed Distributiv Law for Multiplication and Addition.**

371. Theorem. $(\alpha + \beta + \cdots + \kappa)(\lambda + \mu + \cdots + \rho)$

$$= \alpha\lambda + \beta\lambda + \cdots + \kappa\lambda + \alpha\mu + \beta\mu + \cdots + \kappa\mu + \cdots$$
$$+ \alpha\rho + \beta\rho + \cdots + \kappa\rho.$$

This is proved by applying §§ 369 and 370 in succession.

This theorem may be stated: *The product of two complex sums is equal to the complex sum of all the products that can be formd by multiplying an element of the first sum by an element of the second.*

372. We may extend the law of § 371 to the product of two algebraic sums. Change all the $-$ signs befor integers to $+$, at the same time changing the signs over the corresponding integers. Then by § 371 the given product may be written as a sum of products, the only minus signs being those over integers. These may now be removed by §§ 326, 227.

For example $(\alpha - \beta - \gamma)(- \delta + \varepsilon)$

$$= (\alpha + \bar{\beta} + \bar{\gamma})(\bar{\delta} + \varepsilon) = \alpha\bar{\delta} + \alpha\varepsilon + \bar{\beta}\bar{\delta} + \bar{\beta}\varepsilon + \bar{\gamma}\bar{\delta} + \bar{\gamma}\varepsilon$$
$$= - \alpha\delta + \alpha\varepsilon + \beta\delta - \beta\varepsilon + \gamma\delta - \gamma\varepsilon.$$

Thus we hav proved the following theorem:

Theorem. *The product of two algebraic sums is equal to the algebraic sum of all the products that can be formd by multiplying an element of the first sum by an element of the second, the sign befor any one of these products being plus or minus according as the signs befor its two elements in the given sums ar like or unlike.*

373. Definition. When a product of two algebraic sums is changed to an algebraic sum of products as in § 372, these latter products ar calld **partial products** and the original product is said to be **distributed.**

374. Theorem. $an = na$.

For $an = a + a + a + \cdots n$ times § 301.

$= 1.a + 1.a + 1.a + \cdots n$ times §§ 316, 252.

$= (1 + 1 + 1 + \cdots n \text{ times})a$ § 370.

$= na.$ §§ 63, 329.

375. Theorem. $a\beta = \beta a$.

For $a\beta = ab_1 - ab_2$ § 315.

$= b_1a - b_2a$ §§ 374, 239.

$= (b_1 - b_2)a$ § 367.

$= \beta a.$ §§ 240, 329.

This theorem is calld the **Commutativ Law for Multiplication** (1). It may be stated:

In the operation of multiplication the multiplicand and multiplier may be interchanged.

376. Definition. On account of this interchangability it is proper to giv the multiplicand and multiplier a common name. They ar calld the **factors** (2)

(1) For a proof of the commutativ law for multiplication of primary numbers see A. M. Legendre, "Theorie des Nombres," p. 1; Dirichlet-Dedekind, "Vorlesungen über Zahlentheorie," 1879, 4th ed., 1894, p. 2; Tannery, p. 58; Beman and Smith, p. 38.

(2) See Schubert, "Encyklopädie," p. 15.

of the product. The first factor is calld the **prefactor,** the second the **postfactor** ([1]).

377. Theorem. *Every even integer can be written in the form* $2a$; *every odd integer can be written in either of the forms* $2a + 1$ *and* $2a - 1$; *and conversely, every integer that can be written in the form* $2a$ *is even; every integer that can be written in either of the forms* $2a + 1$ *and* $2a - 1$ *is odd.*

In the first place, since $0 = 2.0$, the theorem is true for the integer zero.

Second, every positiv even integer can be obtaind by adding a number of twos to zero. § 294.

If the integer equals $0 + 2 + 2 + 2 + \cdots n$ times, then it equals $2n$. §§ 209, 301.

Hence the theorem is true for every positiv even integer.

Next, every negativ even integer can be obtaind by subtracting a number of twos from zero.

Thus it equals $0 - 2 - 2 - 2 - \cdots m$ times, or $-2m$, which equals $2(-m)$.

Hence the theorem is true for every negativ even integer.

Moreover, since every odd integer can be obtaind by adding one to the even integer preceding it in the natural series or subtracting one from the even integer succeeding it, every odd integer can be written in either of the forms $2a + 1$ and $2a - 1$.

---

(1) See J. W. Gibbs, "Vector Analysis," p. 41.

The converse is proved by reversing the steps.

378. Definition. A complex expression containing integers connected by multiplication signs only, with parentheses to indicate the order in which the successiv operations ar performd, is calld a **complex product.** The integers from which it is formd ar calld its **elements.**

379. Theorem. $(\alpha\beta)n = \alpha(\beta n)$.

For $(\alpha\beta)n = \alpha\beta + \alpha\beta + \alpha\beta + \cdots n$ times § 301.

$= \alpha(\beta + \beta + \beta + \cdots n \text{ times})$ § 369.

$= \alpha(\beta n)$. §§ 301, 330.

380. Theorem. $(\alpha\beta)\gamma = \alpha(\beta\gamma)$.

For $(\alpha\beta)\gamma = (\alpha\beta)c_1 - (\alpha\beta)c_2$ § 315.

$= \alpha(\beta c_1) - \alpha(\beta c_2)$ §§ 379, 239.

$= \alpha(\beta c_1 - \beta c_2)$ § 366.

$= \alpha(\beta\gamma)$. §§ 315, 330.

This theorem is calld the **Associativ Law for Multiplication** (1).

381. Theorem. *A complex product is independent of the arrangement of its parentheses and the order of its elements.*

The proof of this theorem is formally the same as the proof of § 250, the only difference being that multiplication signs replace addition signs.

---

(1) For a proof of the associativ law for multiplication of primary numbers see Legendre, p. 2; Dirichlet-Dedekind, p. 1; Beman and Smith, p. 43.

This theorem is the **Generalized Associativ and Commutativ Law for Multiplication** (1).

382. Hence we may speak of a product $\alpha\beta\gamma\delta\varepsilon\cdots$ without parentheses and may write the elements of the product, α, β, γ, δ, ε, $\cdots$, in any order we please without altering the product.

383. Definition. The elements of a complex product ar also calld its **factors.**

384. Theorem. *The product of the left members of any number of equalities is equal to the product of the right members.*

The proof of this theorem is formally the same as the proof of § 252.

385. Theorem. *The product of any number of positiv integers is a positiv integer; the product of any number of integers, one of which is zero, is a zero integer; the product of an odd number of negativ integers is a negativ integer; the product of an even number of negativ integers is a positiv integer.*

The proof of the first part of the theorem is formally the same as that of § 254.

So also is the proof of the second part, if we write the zero integer first.

To prove the last two parts, we know that the product of two negativ integers is positiv. § 320.

Also, if an integer is multiplied by a negativ integer, the sign of the result is opposit to that of the first integer. § 320.

(1) See Legendre, p. 2.

Hence, when we form a product of any number of negativ integers by first multiplying two together, then their product by a third, this product by a fourth, and so on, the products will be alternately positiv and negativ, being positiv for two integers.

The positiv products will then correspond to an even number of integers and the negativ products to an odd number.

386. Theorem. *If one of a series of integers is zero, their product is equal to zero; and conversely, if the product of a series of integers is equal to zero, one of them must be zero.*

The proof of the direct theorem is formally the same as the proof of § 255, if we write the zero integer first.

The converse is proved by using § 343 repeatedly.

387. Theorem. $(\pm \alpha)(\pm \beta)(\pm \gamma) \cdots = \pm (\alpha\beta\gamma \cdots)$, *the plus sign being used in the right member, if ther ar no minus signs or an even number of minus signs in the left member, the minus sign being used in the right member, if ther ar an odd number of minus signs in the left member.*

This theorem follows from § 326 by a method similar to that used to prove the last two parts of § 385.

388. Theorem. *If one of a series of integers is even, their product is even; if all ar odd, their product is odd; and conversely, if the product of a series of integers is even, one of them must be even; if the product is odd, all must be odd.*

For the proof of the first part of the direct theorem, if the series of integers is $\alpha, \beta, \gamma, \delta, \cdots$, the first of the products $\alpha\beta$, $(\alpha\beta)\gamma$, $((\alpha\beta)\gamma)\delta$, $\cdots$ which has an even integer for one of its factors is even. § 360.

The method of § 254 may then be applied to prove that the product of the whole series is even.

The second part of the direct theorem follows immediately from § 360 by the method of § 254.

The converse theorem may be proved by the method of exhaustion.

389. Theorem. *The product of any number of sums is equal to the sum of all the products that can be formd, each having as a factor one and only one element of each of the given sums* (1).

This may be proved by first applying the theorem of § 371 to the product of the first two sums, then applying the same theorem to the product of this result and the next sum, and so on.

For example, $(\alpha + \beta + \gamma)(\delta + \varepsilon)(\zeta + \eta + \theta)$ $= (\alpha\delta + \beta\delta + \gamma\delta + \alpha\varepsilon + \beta\varepsilon + \gamma\varepsilon)(\zeta + \eta + \theta) = \alpha\delta\zeta + \beta\delta\zeta + \gamma\delta\zeta + \alpha\varepsilon\zeta + \beta\varepsilon\zeta + \gamma\varepsilon\zeta + \alpha\delta\eta + \beta\delta\eta + \gamma\delta\eta + \alpha\varepsilon\eta + \beta\varepsilon\eta + \gamma\varepsilon\eta + \alpha\delta\theta + \beta\delta\theta + \gamma\delta\theta + \alpha\varepsilon\theta + \beta\varepsilon\theta + \gamma\varepsilon\theta$.

390. By the same method as that employed in § 372 we may extend the preceding theorem to the product of any number of algebraic sums. Thus we get the following theorem:

Theorem. *The product of any number of algebraic sums is equal to the algebraic sum of all the products*

(1) See G. Chrystal, "Text-book of Algebra," 1893, Part I., p. 42.

that can be formd each having as a factor one and only one element of each of the given algebraic sums, the sign befor any one of these products being plus, if none of its elements or an even number ar immediately preceded by minus signs in the given sums, and being minus, if an odd number are so preceded (1).

391. By § 326 the product of a positiv and a negativ integer, or of two negativ integers, may be found when the product of the corresponding positiv integers is known.

To multiply two positiv integers, or primary numbers, a and b, we add b a's together. The sum obtaind is the required product.

Or we may join b groups of a objects each so as to form a single group. The number of objects in the group thus obtaind is the required product.

392. Multiplication Table (2). In order to make a table of the products obtaind by multiplying all possible pairs of the numbers 1, 2, 3, ⋯, 10, except those products which may not be represented by the symbols so far defined, we divide a square into compartments as in § 74 and write the numbers 1, 2, 3, ⋯, 10 along the top and left-hand sides of the square as in that article.

In each compartment we wish to place the number which is the product of the two numbers found respectivly at the left of the row and at the top of the column in which the compartment lies.

---

(1) See Chrystal, p. 50. (2) See Tannery, p. 68.

Thus in the first row of compartments we should write the numbers that represent the products 1×1, 1×2, 1×3, $\cdots$, or 1, 2, 3, $\cdots$

In the second row we should write the products 2×1, 2×2, 2×3, $\cdots$ These ar very easily found as follows. We write the natural series of numbers 1, 2, 3, 4, 5, 6, $\cdots$. Then starting at 1 we count to the second number. This is 2×1. Counting to the second number beyond this, the number arrived at is 2×2 (§ 73). Counting to the second place beyond this, we find 2×3, and so on. Thus in the second row we write the numbers 2, 4, 6, $\cdots$

The numbers to be written in the third row ar the multiples of 3, 3×1, 3×2, 3×3, $\cdots$, which ar found from the natural series by counting by threes. And so on.

It will be noticed that the table thus formd, given below, is symmetrical with reference to its principal diagonal, which might hav been forseen from the commutativ law for multiplication.

MULTIPLICATION TABLE.

	1	2	3	4	5	6	7	8	9	ϒ	∩	ℭ	ƨ	Ұ	Ꞓ	10
1	1	2	3	4	5	6	7	8	9	ϒ	∩	ℭ	ƨ	Ұ	Ꞓ	10
2	2	4	6	8	ϒ	ℭ	Ұ	10								
3	3	6	9	ℭ	Ꞓ											
4	4	8	ℭ	10												
5	5	ϒ	Ꞓ													
6	6	ℭ														
7	7	Ұ														
8	8	10														
9	9															
ϒ	ϒ															
∩	∩															
ℭ	ℭ															
ƨ	ƨ															
Ұ	Ұ															
Ꞓ	Ꞓ															
10	10															

CHAPTER V.

NUMERICAL VALUE.

393. Definition. The **numerical value, absolute value,** or **modulus** of an integer is, if the integer is positiv or zero, the integer itself; if negativ, its opposit.

We will use the symbol $|a|$ to denote the numerical value (¹) of a. This symbol is univalent (§§ 161, 153).

Thus $|5| = 5$, $|0| = 0$, $|\bar{3}| = 3$.

394. Theorem. *The numerical value of a positiv or negativ integer is positiv; the numerical value of a zero integer is zero.*

395. Theorem. $|\bar{a}| = |a|$.

If a is positiv, $\bar{a}$ is negativ. § 164.

Then $|a| = a$ and $|\bar{a}| = a$. § 393.

Therfor $|\bar{a}| = |a|$. § 173.

If a is zero, $\bar{a}$ is zero.

Then $|a| = 0$ and $|\bar{a}| = 0$.

Therfor $|\bar{a}| = |a|$.

If a is negativ, $\bar{a}$ is positiv.

Then $|a| = \bar{a}$ and $|\bar{a}| = \bar{a}$.

Therfor $|\bar{a}| = |a|$.

396. Theorem. *If two integers ar equal, their numerical values ar equal.*

(¹) The symbol $|a|$ is due to K. Weierstrass. It was used by him in 1841. See his works, Vol. I, p. 67.

Let $\alpha = \beta$. Then, if α is positiv, β is positiv. § 176.
Hence $|\alpha| = \alpha$ and $|\beta| = \beta$.
Therfor $|\alpha| = |\beta|$.
Similarly the other cases ar proved.

397. Theorem. *If* $\alpha = \bar{\beta}$, $|\alpha| = |\beta|$.

For, by § 396, $|\alpha| = |\bar{\beta}|$. But $|\bar{\beta}| = |\beta|$. § 395.

398. Theorem. *If* $|\alpha| = 0$, $\alpha = 0$.

Prove by the method of exhaustion.

399. Theorem. *If* $|\alpha| = |\beta|$, *either* $\alpha = \beta$ *or* $\alpha = \bar{\beta}$.

If $\beta = 0$, $|\beta| = 0$. Hence $|\alpha| = 0$ and $\alpha = 0$.
Therfor $\alpha = \beta$.
Similarly, if $\alpha = 0$, $\alpha = \beta$.
Ther remain four cases to be considered:
1st, when α and β ar both positiv;
2d, when α is positiv and β negativ;
3d, when α is negativ and β positiv;
4th, when α and β are both negativ.

Case 1. Here $|\alpha| = \alpha$ and $|\beta| = \beta$.
Hence, since $|\alpha| = |\beta|$, $\alpha = \beta$.

Case 2. Here $|\alpha| = \alpha$ and $|\beta| = \bar{\beta}$.
Hence $\alpha = \bar{\beta}$.

Case 3. Here $\alpha = \beta$.
Therfor $\alpha = \bar{\beta}$.

Case 4. Here $\bar{\alpha} = \bar{\beta}$.
Therfor $\alpha = \beta$.

In all cases, therfor, either $\alpha = \beta$ or $\alpha = \bar{\beta}$.

400. Theorem. *The numerical value of the sum of two integers is not greater than the sum of their numerical values.*

Case 1. If α and β ar both positiv, $|\alpha| = \alpha$ and $|\beta| = \beta$.

Then $\alpha + \beta$ is positiv (§ 206) and $|\alpha + \beta| = \alpha + \beta$.

Therfor $|\alpha + \beta| = |\alpha| + |\beta|$.

Case 2. If either α or β is zero, its numerical value is zero and $\alpha + \beta$ is equal to the other integer.

Hence $|\alpha + \beta|$ and $|\alpha| + |\beta|$ ar both equal to the numerical value of the other integer.

Therfor $|\alpha + \beta| = |\alpha| + |\beta|$.

Case 3. If α and β ar both negativ, $|\alpha| = \bar{\alpha}$ and $|\beta| = \bar{\beta}$.

Then $\alpha + \beta$ is negativ and $|\alpha + \beta| = -(\alpha + \beta)$.

Moreover $|\alpha| + |\beta| = \bar{\alpha} + \bar{\beta} = -(\alpha + \beta)$.

Therfor $|\alpha + \beta| = |\alpha| + |\beta|$.

Case 4. If α is positiv and β negativ $|\alpha| = \alpha$ and $|\beta| = \bar{\beta}$.

Then $\alpha + \beta = \alpha - \bar{\beta} = |\alpha| - |\beta|$.

Now, if $|\alpha| > |\beta|$, $|\alpha| - |\beta|$ is positiv and therfor $\alpha + \beta$ is positiv.

Hence $|\alpha + \beta| = \alpha + \beta = |\alpha| - |\beta|$.

But $|\alpha| - |\beta| < |\alpha| + |\beta|$. § 235.

Therfor $|\alpha + \beta| < |\alpha| + |\beta|$.

If $|\alpha| = |\beta|$, $|\alpha| - |\beta|$ is zero and therfor $\alpha + \beta$ is zero.

Hence $|\alpha + \beta| = \alpha + \beta = |\alpha| - |\beta|$.

Therfor $|\alpha + \beta| < |\alpha| + |\beta|$.

If $|\alpha| < |\beta|$, $|\alpha| - |\beta|$ is negativ and therfor $\alpha + \beta$ is negativ.

Hence $|\alpha + \beta| = -(\alpha + \beta) = |\beta| - |\alpha|$.

But $|\beta| - |\alpha| < |\alpha| + |\beta|$.

Therfor $|\alpha + \beta| < |\alpha| + |\beta|$.

Similarly we may prove that, if α is negativ and β positiv, $|\alpha + \beta| < |\alpha| + |\beta|$.

Therfor, in all cases, $|\alpha + \beta| \leqq |\alpha| + |\beta|$.

That is, $|\alpha + \beta| \not> |\alpha| + |\beta|$.

401. Theorem. *The numerical value of the sum of any number of integers is not greater than the sum of their numerical values* (1).

402. Theorem. *The numerical value of the difference of two integers is not greater than the sum of their numerical values.*

For $\alpha - \beta = \alpha + \bar{\beta}$.

Hence $|\alpha - \beta| = |\alpha + \bar{\beta}| \not> |\alpha| + |\bar{\beta}| = |\alpha| + |\beta|$.

Therfor $|\alpha - \beta| \not> |\alpha| + |\beta|$.

403. Theorem. *The numerical value of any algebraic sum is not greater than the sum of the numerical values of its elements.*

404. Theorem. *The numerical value of the product of two integers is the product of their numerical values.*

If α and β ar both positiv, $|\alpha| = \alpha$ and $|\beta| = \beta$.

Then $\alpha\beta$ is positiv (§ 320) and $|\alpha\beta| = \alpha\beta$.

Therfor $|\alpha\beta| = |\alpha||\beta|$.

If either α or β is zero, its numerical value is zero and $\alpha\beta$ is zero.

Hence $|\alpha\beta|$ and $|\alpha||\beta|$ ar both zero.

Therfor $|\alpha\beta| = |\alpha||\beta|$.

If α and β are both negativ, $|\alpha| = \bar{\alpha}$ and $|\beta| = \bar{\beta}$.

(1) See Stolz und Gmeiner, p. 64.

Then $\alpha\beta$ is positiv and $|\alpha\beta| = \alpha\beta$.

Moreover $|\alpha||\beta| = \bar{\alpha}\bar{\beta} = \alpha\beta$.

Therfor $|\alpha\beta| = |\alpha||\beta|$.

If α is positiv and β negativ, $|\alpha| = \alpha$ and $|\beta| = \bar{\beta}$.

Then $\alpha\beta$ is negativ and $|\alpha\beta| = -(\alpha\beta)$.

Therfor $|\alpha\beta| = |\alpha||\beta|$.

Similarly we may prove the remaining case when α is negativ and β positiv.

405. Theorem. *The numerical value of the product of any number of integers is equal to the product of the numerical values of those integers.*

406. Theorem. *If* $|\alpha| > 1$ *and* $\beta \neq 0$, $|\alpha\beta| > |\beta|$.

For, if $|\alpha| > 1$ and $\beta \neq 0$, $|\alpha||\beta| > 1 \times |\beta|$. §332.

Or $|\alpha\beta| > |\beta|$. § 404.

407. Theorem. $|\alpha\alpha| = \alpha\alpha$.

For $\alpha\alpha$ is either zero or positiv. § 320.

408. Theorem. *If* $|\alpha\beta| < \gamma\gamma$, *either* α *or* β *is numerically less than* γ.

For, if neither α nor β were numerically less than γ, we would have $|\alpha| \geqq |\gamma|$ and $|\beta| \geqq |\gamma|$.

Hence $|\alpha||\beta| \geqq |\gamma||\gamma|$. §§ 331, 349, 350.

Or $|\alpha\beta| \geqq \gamma\gamma$. §§ 404, 407.

CHAPTER VI.

DIVISION.

DIVISIBILITY. FACTORS. QUOTIENT.

409. Definition. Having given two integers α and β, if β is not zero and ther is some third integer φ such that $\alpha = \varphi\beta$, α is said to be **divisible by** β (1); if α is not zero and ther is some integer φ such that $\varphi\alpha = \beta$, α is said to be a **divisor of** β.

Thus, since $8 = 4 \times 2$, 8 is divisible by 2; since $4 \times 2 = 8$, 2 is a divisor of 8.

The reasons for excluding the cases wher β and α ar respectivly zero will appear in §§ 422, 424.

The statement "α is divisible by β" we will write $\alpha \gtrdot \beta$; the statement "α is a divisor of β" $\alpha \lessdot \beta$.

The definitions of the terms "divisible by" and "divisor of" may, then, be stated in symbols:

If $\alpha = \varphi\beta$, wher β is not zero, then $\alpha \gtrdot \beta$; and conversely, if $\alpha \gtrdot \beta$, then β is not zero and $\alpha = \varphi\beta$.

If $\varphi\alpha = \beta$, wher α is not zero, then $\alpha \lessdot \beta$; and conversely, if $\alpha \lessdot \beta$, then α is not zero and $\varphi\alpha = \beta$.

The statements $\alpha \gtrdot \beta$ and $\alpha \lessdot \beta$ may be calld **divisibilities.**

---

(1) See Dirichlet-Dedekind, p. 5.

410. It should be noticed that, if β is zero and $\alpha = \varphi\beta$, then α is zero. Similarly, if α is zero and $\varphi\alpha = \beta$, β is zero.

Hence by the definitions of § 409 none of the statements $0 \gtrdot 0$, $0 \lessdot 0$, $\alpha \gtrdot 0$, $0 \lessdot \alpha$ is true.

411. Theorem. *If* $\alpha = \varphi\beta$ *and* $\varphi \mp \chi$, *then* $\alpha \mp \chi\beta$, *provided* β *is not zero.*

For, since $\varphi \mp \chi$ and β is not zero, $\varphi\beta \mp \chi\beta$. § 338.

Hence, since $\alpha = \varphi\beta$, $\alpha \mp \chi\beta$. § 174.

412. Definition. From the last theorem we see that if $\alpha = \varphi\beta$, wher β is not zero, ther is no integer χ, different from φ, such that $\alpha = \chi\beta$.

Hence, if α and β are given, β not being zero, φ is determind by the equality $\alpha = \varphi\beta$.

This passiv number φ, which depends only on α and β and is such that $\alpha = \varphi\beta$, we will call the **right-handed quotient** (1) of α and β.

This quotient we will represent by the symbol $\alpha : \beta$ (2), which is read "α divided by β."

The quotient $\alpha : \beta$ is, then, univalent (3), when it exists, that is, when β is not zero and ther is some third integer φ, such that $\alpha = \varphi\beta$.

The number α may be calld the **first element** of the quotient and β the **second element.**

413. Theorem. *If the right-handed quotient* $\alpha : \beta$ *exists and is the integer* φ, *then* $\alpha : \beta = \varphi$.

(1) See Schubert, "Encyklopädie," p. 16.

(2) See Schubert, "Encyklopädie," p. 16; Stolz und Gmeiner, p. 25. The symbol : to denote a quotient was used by A. C. Clairaut in 1760.

(3) See Stolz und Gmeiner, p. 25.

This follows from § 170.

414. Theorem. *If $\alpha = \varphi\beta$, wher β is not zero, then $\alpha : \beta = \varphi$; and conversely, if $\alpha : \beta = \varphi$, then β is not zero and $\alpha = \varphi\beta$.*

This follows immediately from the definition of right-handed quotient and § 413.

415. The symbols $0 : 0$, $\alpha : 0$ (1) ar both as yet meaningless.

416. Definition. If the restriction in § 412 that β should not be zero had been omitted, that article would not hav defined the symbol $\alpha : \beta$.

For, if β were zero, α would also be zero and α would equal $\varphi\beta$ whatever integer φ might be.

The symbol $\alpha : \beta$, α and β being both zero, would then represent any integer. It would not be defined, made definit.

A symbol like $0 : 0$, which, tho compounded of two definit numbers, would represent more than one number, is calld a **multivalent symbol** (2).

The symbol $0 : 0$ would in fact represent not only more than one integer, but any integer. Such a symbol, one representing an unlimited number of numbers, is calld an **indeterminate** symbol.

Multivalent symbols, if used at all, must be used with great care, as the laws which hold for univalent symbols do not apply to them.

417. Theorem. *If $\varphi\alpha = \beta$ and $\varphi \neq \chi$, then $\chi\alpha \neq \beta$, provided α is not zero.*

---

(1) See Schubert, "Encyklopädie," p. 17.

(2) See Schubert, "Encyklopädie," p. 17.

418. From this theorem we see that the equality $\varphi\alpha = \beta$ determins the integer φ, provided α is not zero. We will call the integer φ the **right-handed ratio** of β to α, which we will represent by the symbol β/α, which may be read "β over α."

The ratio β/α is, then, univalent, when it exists, that is, when α is not zero and ther is some third integer φ, such that $\varphi\alpha = \beta$.

419. Theorem. *If the right-handed ratio β/α exists and is the integer φ, then $\varphi = \beta/\alpha$.*

420. Theorem. *If $\varphi\alpha = \beta$, wher α is not zero, then $\varphi = \beta/\alpha$; and conversely, if $\varphi = \beta/\alpha$, then α is not zero and $\varphi\alpha = \beta$.*

This follows immediately from the definition of right-handed ratio and § 419.

421. We will now derive some theorems connecting the ideas represented by the symbols $\dot{>}$, $\dot{<}$, : , /, which hav been independently defined.

422. Theorem. *If $\alpha \dot{>} \beta$, the right-handed quotient $\alpha : \beta$ exists, and conversely.*

The proof of this theorem is formally the same as that of § 86.

The reason is now evident why in § 409 we excluded the case of β being zero. If we had not done so, we should hav $0 \dot{>} 0$, but the right-handed quotient $0 : 0$ not existing.

423. From the last theorem it follows that the symbol $\alpha : \beta$ as yet has no meaning, unless $\alpha \dot{>} \beta$.

424. Theorem. *If* $\alpha \lessdot \beta$, *the right-handed ratio* β / α *exists, and conversely.*

The reason is now evident why in § 409 we excluded the case of α being zero.

425. Theorem. *If* $\alpha \gtrdot \beta$, *then* $\beta \lessdot \alpha$ *and* $\alpha / \beta = \alpha : \beta$; *and conversely, if* $\beta \lessdot \alpha$, *then* $\alpha \gtrdot \beta$ *and* $\alpha : \beta = \alpha / \beta$.

This is proved in the same way as § 89.

The proof depends on the commutativ law for equality of integers.

426. We may therfor, whenever we hav either of the statements $\alpha \gtrdot \beta$, $\beta \lessdot \alpha$, replace it by the other and may replace either of the expressions $\alpha : \beta$, α / β by the other.

We will find it convenient in future seldom to use the symbol /, but to use in its stead : .

The symbols $\gtrdot$ and $\lessdot$ will both be useful, tho we will employ chiefly $\gtrdot$. We should always bear in mind that the statement $\alpha \gtrdot \beta$ may at pleasure be replaced by the statement $\beta \lessdot \alpha$.

427. Definition. Having given two integers α and β, if β is not zero and ther is some third integer φ such that $\alpha = \beta\varphi$, α is said to be a **multiple of** β; if α is not zero and ther is some integer φ such that $\alpha\varphi = \beta$, α is said to be a **factor of** β.

The statement "α is a multiple of β" may be written $\alpha \gtrdot \beta$; the statement "α is a factor of β" $\alpha \lessdot \beta$.

The definitions of the terms "multiple of" and "factor of" may, then, be stated in symbols :

If $\alpha = \beta\varphi$, wher β is not zero, then $\alpha \gtrdot \beta$, and conversely.

If $\alpha\varphi = \beta$, wher α is not zero, then $\alpha \lessdot \beta$, and conversely.

428. Theorem. *If* $\alpha = \beta\varphi$ *and* $\varphi \neq \chi$, *then* $\alpha \neq \beta\chi$, *provided* β *is not zero.*

429. Definition. From this theorem we see that the equality $\alpha = \beta\varphi$ determins the integer φ, provided β is not zero.

This activ number φ, which depends only on β and α and is such that $\alpha = \beta\varphi$, we will call the **left-handed quotient** (1) of β and α. For this quotient we will use the symbol $\beta \cdot\cdot\, \alpha$, which may be read "$\beta$ divided into α."

The quotient $\beta \cdot\cdot\, \alpha$ is, then, univalent, when it exists, that is, when β is not zero and ther is a third integer φ, such that $\alpha = \beta\varphi$.

430. Theorem. *If the left-handed quotient* $\beta \cdot\cdot\, \alpha$ *exists and is the integer* φ, *then* $\beta \cdot\cdot\, \alpha = \varphi$.

431. Theorem. *If* $\alpha = \beta\varphi$, *wher* β *is not zero, then* $\beta \cdot\cdot\, \alpha = \varphi$, *and conversely.*

432. Theorem. *If* $\alpha\varphi = \beta$ *and* $\varphi \neq \chi$, *then* $\alpha\chi \neq \beta$, *provided* β *is not zero.*

433. Definition. From this theorem we see that the equality $\alpha\varphi = \beta$ determins the integer φ, provided α is not zero. We will call this integer the **left-handed ratio** of α to β. For this ratio we will use the symbol $\alpha\backslash\beta$ (2), which may be read "α under β."

(1) See foot-note p. 36.

(2) This symbol seems to be due to A. Cayley.

The ratio $\alpha\backslash\beta$ is, then, univalent, when it exists, that is, when α is not zero and ther is a third integer φ, such that $\alpha\varphi = \beta$.

434. Theorem. *If the left-handed ratio* $\alpha\backslash\beta$ *exists and is the integer* φ, *then* $\varphi = \alpha\backslash\beta$.

435. Theorem. *If* $\alpha\varphi = \beta$, *wher* α *is not zero, then* $\varphi = \alpha\backslash\beta$, *and conversely.*

436. Theorem. *If* $\alpha \gtrdot \beta$, *the left-handed quotient* $\beta \cdot\cdot \alpha$ *exists, and conversely.*

437. Theorem. *If* $\alpha \lessdot \beta$, *the left-handed ratio* $\alpha\backslash\beta$ *exists, and conversely.*

438. Theorem. *If* $\alpha \gtrdot \beta$, *then* $\alpha \gtrdot \beta$ *and* $\beta \cdot\cdot \alpha = \alpha : \beta$; *and conversely, if* $\alpha \gtrdot \beta$, *then* $\alpha \gtrdot \beta$ *and* $\alpha : \beta = \beta \cdot\cdot \alpha$.

The proof of this theorem depends on the commutativ law for multiplication.

439. Theorem. *If* $\alpha \lessdot \beta$, *then* $\alpha \lessdot \beta$ *and* $\alpha\backslash\beta = \beta/\alpha$; *and conversely, if* $\alpha \lessdot \beta$, *then* $\alpha \lessdot \beta$ *and* $\beta/\alpha = \alpha\backslash\beta$.

The proof of this theorem also depends on the commutativ law for multiplication.

440. Definition. In virtue of the last two theorems we will in future never use the symbols $\gtrdot$ and $\lessdot$, but will use in their stead $\gtrdot$ and $\lessdot$ respectivly. Similarly instead of $\beta \cdot\cdot \alpha$ and $\beta\backslash\alpha$ it is customary to use $\alpha : \beta$ and α/β.

In virtue of §§ 425, 438, 439 we will in future instead of the four terms "right-handed quotient," "right-handed ratio," "left-handed quotient," and "left-handed ratio" use the single term "quotient."

Besides the symbols already mentiond for the quotient of α and β the symbols $\alpha \div \beta$ and $\frac{\alpha}{\beta}$ ar in common use [1].

441. Definition. The symbols $\not\gtrdot$ and $\not\lessdot$ may be used to mean "is not divisible by" and "is not a divisor of" respectivly.

442. Theorem. *Every integer except zero is divisible by itself, the quotient being unity.*

In symbols $\alpha \gtrdot \alpha$ and $\alpha : \alpha = 1$, if $\alpha \neq 0$.

Prove from §§ 316, 409, 414.

443. Theorem. *Every integer is divisible by unity, the quotient being the integer itself.*

In symbols $\alpha \gtrdot 1$ and $\alpha : 1 = \alpha$.

444. Theorem. *Zero is divisible by every integer except itself, the quotient being zero.*

In symbols $0 \gtrdot \alpha$ and $0 : \alpha = 0$, if $\alpha \neq 0$.

445. Theorem. *If* $\alpha \gtrdot \beta$ *and* $\alpha : \beta = 0$, *then* $\alpha = 0$.

For, if $\alpha \gtrdot \beta$ and $\alpha : \beta = 0$, $\alpha = 0 \cdot \beta = 0$.

446. Theorem. It follows from the last two theorems that *the quotient of two integers is zero when, and only when, its first element is zero.*

447. Theorem. *If* $\alpha = \beta$, *then* $\alpha \gtrdot \beta$ *and* $\alpha : \beta = 1$, *provided* $\beta \neq 0$.

---

(1) The symbol — for a quotient is found in the "Liber Abaci" of Leonardo of Pisa. This symbol in both the forms $\frac{\alpha}{\beta}$, α/β is probably due to the Hindoos. The symbol $\div$ was probably originally formd as a combination of — and : .

For, if $\alpha = \beta$, $\alpha = 1 \cdot \beta$. Hence $\alpha \gtrdot \beta$ and $\alpha : \beta = 1$.

448. Theorem. *Every even integer is divisible by two.*

For, if β is any even integer, $\beta = 2\alpha$. § 377.

449. Theorem. *No odd integer is divisible by an even integer.*

For, if $\alpha \gtrdot \beta$, $\alpha = \varphi\,\beta$.

Hence, if α is odd, φ and β must both be odd. § 360.

450. Theorem. *If an even integer is divisible by an odd integer, the quotient is even.*

For, if $\alpha \gtrdot \beta$, $\alpha = \varphi\beta$ and $\alpha : \beta = \varphi$.

Hence, if α is even, either φ or β must be even. § 360.

451. Theorem. *If an odd integer is divisible by an odd integer, the quotient is odd.*

452. The quotient of two integers may be connected with other integers by signs for sum, difference, product, and quotient, provided that the first element of every quotient is divisible by the second element. Thus we may hav expressions like

$$((\alpha : \beta) \times \gamma) : (\delta : \varepsilon)$$

or $$((\alpha - \beta) : \gamma) \times (\delta + \varepsilon) + (\zeta : (\eta - \theta + \iota)).$$

Every such expression is univalent, since every sum, difference, product, and quotient in the expression is univalent.

Generally in complex expressions involving quotients with sums or differences, or both, parentheses ar omitted from around quotients. They should never be omitted from around sums or differences, when such omission would cause misunderstanding.

Thus $\alpha : \beta + \gamma : \delta$ is understood to mean $(\alpha : \beta) + (\gamma : \delta)$ and not $(\alpha : (\beta + \gamma)) : \delta$.

453. Theorem. *$\alpha \times \beta \mathrel{\dot{>}} \beta$ and $(\alpha \times \beta) : \beta = \alpha$, provided β is not zero.*

For $\alpha \times \beta = \alpha \times \beta$.

Therfor $\alpha \times \beta \mathrel{\dot{>}} \beta$ and $(\alpha \times \beta) : \beta = \alpha$.

454. Theorem. *A product is divisible by each of its factors, provided that factor is not zero.*

455. Theorem. *If $\alpha \mathrel{\dot{>}} \beta$, then $(\alpha : \beta) \times \beta = \alpha$.*

This is proved in the same way as § 110.

456. Theorem. *If $\alpha \mathrel{\dot{>}} \beta$, then $\alpha \times \gamma \mathrel{\dot{>}} \beta \times \gamma$ and $(\alpha \times \gamma) : (\beta \times \gamma) = \alpha : \beta$, provided γ is not zero; and conversely, if $\alpha \times \gamma \mathrel{\dot{>}} \beta \times \gamma$, then $\alpha \mathrel{\dot{>}} \beta$ and $\alpha : \beta = (\alpha \times \gamma) : (\beta \times \gamma)$.*

For the direct theorem, since $\alpha \mathrel{\dot{>}} \beta$,

$$\alpha = \varphi\beta \quad \text{and} \quad \alpha : \beta = \varphi.$$

Hence $\alpha\gamma = (\varphi\beta)\gamma = \varphi(\beta\gamma)$ and $(\alpha\gamma) : (\beta\gamma) = \varphi$.

Therfor $\alpha\gamma \mathrel{\dot{>}} \beta\gamma$ and $(\alpha\gamma) : (\beta\gamma) = \alpha : \beta$.

For the converse theorem, since $\alpha\gamma \mathrel{\dot{>}} \beta\gamma$,

$$\alpha\gamma = \varphi(\beta\gamma) = (\varphi\beta)\gamma \quad \text{and} \quad (\alpha\gamma) : (\beta\gamma) = \varphi.$$

Hence $\alpha = \varphi\beta$ and $\alpha : \beta = \varphi$.

Therfor $\alpha \mathrel{\dot{>}} \beta$ and $\alpha : \beta = (\alpha\gamma) : (\beta\gamma)$.

457. Theorem. *If $\alpha \mathrel{\dot{>}} \beta$, then $\gamma \times \alpha \mathrel{\dot{>}} \gamma \times \beta$ and $(\gamma \times \alpha) : (\gamma \times \beta) = \alpha : \beta$, provided γ is not zero; and conversely, if $\gamma \times \alpha \mathrel{\dot{>}} \gamma \times \beta$, then $\alpha \mathrel{\dot{>}} \beta$ and $\alpha : \beta = (\gamma \times \alpha) : (\gamma \times \beta)$.*

458. Theorem. *If* $\alpha = \beta$ *and* $\beta \gtrdot \gamma$, *then* $\alpha \gtrdot \gamma$ *and* $\alpha : \gamma = \beta : \gamma$; *and conversely, if* $\alpha \gtrdot \gamma$, $\beta \gtrdot \gamma$, *and* $\alpha : \gamma = \beta : \gamma$, *then* $\alpha = \beta$.

For the direct theorem,

since $\beta \gtrdot \gamma$, $\beta = \varphi\gamma$ and $\beta : \gamma = \varphi$.

Hence, since $\alpha = \beta$, $\alpha = \varphi\gamma$ and $\alpha : \gamma = \varphi$.

Therfor $\alpha \gtrdot \gamma$ and $\alpha : \gamma = \beta : \gamma$.

For the converse theorem, let $\alpha : \gamma = \varphi$.

Then $\beta : \gamma = \varphi$.

Hence $\alpha = \varphi\gamma$ and $\beta = \varphi\gamma$.

Therfor $\alpha = \beta$.

459. Theorem. *If* $\alpha \gtrdot \beta$ *and* $\beta = \gamma$, *then* $\alpha \gtrdot \gamma$ *and* $\alpha : \beta = \alpha : \gamma$; *and conversely, if* $\alpha \gtrdot \beta$, $\alpha \gtrdot \gamma$, *and* $\alpha : \beta = \alpha : \gamma$, *then* $\beta = \gamma$, *provided* α *is not zero.*

For the direct theorem,

since $\alpha \gtrdot \beta$, $\alpha = \varphi\beta$ and $\alpha : \beta = \varphi$.

Since $\beta = \gamma$, $\varphi\beta = \varphi\gamma$.

Hence $\alpha = \varphi\gamma$ and $\alpha : \gamma = \varphi$.

Therfor $\alpha \gtrdot \gamma$ and $\alpha : \beta = \alpha : \gamma$.

For the converse theorem, let $\alpha : \beta = \varphi$.

Then $\alpha : \gamma = \varphi$.

Hence $\alpha = \varphi\beta$ and $\alpha = \varphi\gamma$.

Hence $\varphi\beta = \varphi\gamma$.

Moreover, since α is not zero, $\alpha : \beta$, or φ, is not zero.

Therfor $\beta = \gamma$.

460. Theorem. *If* $\alpha \gtrdot \beta$ *and* $\beta \gtrdot \gamma$, *then* $\alpha \gtrdot \gamma$ *and* $\alpha : \gamma = (\alpha : \beta) \times (\beta : \gamma)$(1).

Since $\alpha \gtrdot \beta$, $\alpha = \varphi\beta$ and $\alpha : \beta = \varphi$.

---

(1) See Tannery, p. 98; Dirichlet-Dedekind, p. 5.

Since $\beta \gtrdot \gamma$, $\beta = \chi\gamma$ and $\beta : \gamma = \chi$.
Hence $\varphi\beta = \varphi(\chi\gamma)$.
Hence $\alpha = (\varphi\chi)\gamma$ and $\alpha : \gamma = \varphi\chi$.
Therfor $\alpha \gtrdot \gamma$ and $\alpha : \gamma = (\alpha : \beta) \times (\beta : \gamma)$.

461. Theorem. *If we hav a series of divisibilities of the same kind, as* $\alpha \gtrdot \beta \gtrdot \gamma \gtrdot \delta \gtrdot \varepsilon \gtrdot \cdots$, *any integer in the series is divisible by any following integer.*

The proof of this theorem is identical in form with that of § 196.

462. Theorem. *If we hav a series of equalities or divisibilities of the form* $\alpha \underset{\gtrdot}{=} \beta \underset{\gtrdot}{=} \gamma \underset{\gtrdot}{=} \delta \underset{\gtrdot}{=} \varepsilon \underset{\gtrdot}{=} \cdots$, *any integer in the series is divisible by any following integer.*

463. Theorem. If $\alpha \gtrdot \beta$ *and* $\beta \gtrdot \gamma$, *then* $\alpha : \gamma \gtrdot \beta : \gamma$ *and* $(\alpha : \gamma) : (\beta : \gamma) = \alpha : \beta$; *and conversely, if* $\alpha \gtrdot \gamma$, $\beta \gtrdot \gamma$, *and* $\alpha : \gamma \gtrdot \beta : \gamma$, *then* $\alpha \gtrdot \beta$ *and* $\alpha : \beta = (\alpha : \gamma) : (\beta : \gamma)$.

For the direct theorem, since $\alpha \gtrdot \beta \gtrdot \gamma$, $\alpha \gtrdot \gamma$ and $\alpha : \gamma = (\alpha : \beta) \times (\beta : \gamma)$.

Therfor $\alpha : \gamma \gtrdot \beta : \gamma$ and $(\alpha : \gamma) : (\beta : \gamma) = \alpha : \beta$.

For the converse theorem,
since $\alpha \gtrdot \gamma$, $\alpha = \varphi\gamma$ and $\alpha : \gamma = \varphi$.
Since $\beta \gtrdot \gamma$, $\beta = \chi\gamma$ and $\beta : \gamma = \chi$.
Since $\alpha : \gamma \gtrdot \beta : \gamma$, $\varphi \gtrdot \chi$.
Hence $\varphi\gamma \gtrdot \chi\gamma$.
Or $\alpha \gtrdot \beta$.

Therfor, by the direct theorem, since $\alpha \gtrdot \beta \gtrdot \gamma$,

$$\alpha : \beta = (\alpha : \gamma) : (\beta : \gamma).$$

464. Theorem. *If* $\alpha \mathrel{\dot{>}} \beta$ *and* $\beta \mathrel{\dot{>}} \gamma$, *then* $\alpha : \gamma \mathrel{\dot{>}} \alpha : \beta$ *and* $(\alpha : \gamma) : (\alpha : \beta) = \beta : \gamma$, *provided* α *is not zero; and conversely, if* $\alpha \mathrel{\dot{>}} \gamma$, $\alpha \mathrel{\dot{>}} \beta$, *and* $\alpha : \gamma \mathrel{\dot{>}} \alpha : \beta$, *then* $\beta \mathrel{\dot{>}} \gamma$ *and* $\beta : \gamma = (\alpha : \gamma) : (\alpha : \beta)$.

465. Theorem. *If* $\alpha \mathrel{\dot{>}} \beta$ *and* $\gamma = \delta$, *then* $\alpha\gamma \mathrel{\dot{>}} \beta\delta$ *and* $(\alpha\gamma) : (\beta\delta) = \alpha : \beta$, *provided* γ *is not zero; and conversely, if* $\alpha\gamma \mathrel{\dot{>}} \beta\delta$ *and* $\gamma = \delta$, *then* $\alpha \mathrel{\dot{>}} \beta$ *and* $\alpha : \beta = (\alpha\gamma) : (\beta\delta)$.

For the direct theorem,

since $\alpha \mathrel{\dot{>}} \beta$, $\alpha\gamma \mathrel{\dot{>}} \beta\gamma$ and $(\alpha\gamma) : (\beta\gamma) = \alpha : \beta$.

Since $\gamma = \delta$, $\beta\gamma = \beta\delta$.

Therfor $\alpha\gamma \mathrel{\dot{>}} \beta\delta$ and $(\alpha\gamma) : (\beta\gamma) = (\alpha\gamma) : (\beta\delta)$.

Also $(\alpha\gamma) : (\beta\delta) = \alpha : \beta$.

For the converse theorem,

since $\gamma = \delta$, $\beta\gamma = \beta\delta$.

Hence,

since $\alpha\gamma \mathrel{\dot{>}} \beta\delta$, $\alpha\gamma \mathrel{\dot{>}} \beta\gamma$ and $(\alpha\gamma) : (\beta\gamma) = (\alpha\gamma) : (\beta\delta)$.

Therfor $\alpha \mathrel{\dot{>}} \beta$ and $\alpha : \beta = (\alpha\gamma) : (\beta\gamma)$.

Also $\alpha : \beta = (\alpha\gamma) : (\beta\delta)$.

466. Theorem. *If* $\alpha \mathrel{\dot{>}} \beta$ *and* $\gamma \mathrel{\dot{>}} \delta$, *then* $\alpha\gamma \mathrel{\dot{>}} \beta\delta$ *and* $(\alpha\gamma) : (\beta\delta) = (\alpha : \beta) \times (\gamma : \delta)$.

If $\gamma = 0$, $\alpha\gamma = 0$ and $\alpha\gamma \mathrel{\dot{>}} \beta\delta$.

Also $\gamma : \delta = 0$ and $(\alpha\gamma) : (\beta\delta) = 0$.

Therfor $(\alpha\gamma) : (\beta\delta) = (\alpha : \beta) \times (\gamma : \delta)$.

If $\gamma \neq 0$, since $\alpha \mathrel{\dot{>}} \beta$, $\alpha\gamma \mathrel{\dot{>}} \beta\gamma$ and $(\alpha\gamma) : (\beta\gamma) = \alpha : \beta$.

Since $\gamma \mathrel{\dot{>}} \delta$, $\beta\gamma \mathrel{\dot{>}} \beta\delta$ and $(\beta\gamma) : (\beta\delta) = \gamma : \delta$.

Therfor $\alpha\gamma \mathrel{\dot{>}} \beta\delta$

and
$$(\alpha\gamma) : (\beta\delta) = [(\alpha\gamma) : (\beta\gamma)] \times [(\beta\gamma) : (\beta\delta)]$$
$$= (\alpha : \beta) \times (\gamma : \delta). \qquad \S\S\ 460, 331.$$

467. Theorem. *If* $\alpha \gtrdot \beta$, *then* $\alpha\gamma \gtrdot \beta$ *and* $(\alpha\gamma) : \beta = (\alpha : \beta) \times \gamma$.

This follows immediately from § 466 by supposing $\delta = 1$.

468. Theorem. *If* $\alpha \gtrdot \beta$, *then* $\gamma\alpha \gtrdot \beta$ *and* $(\gamma\alpha) : \beta = \gamma \times (\alpha : \beta)$.

469. Theorem. *If* $\alpha = \beta$, $\gamma = \delta$, *and* $\beta \gtrdot \gamma$, *then* $\alpha \gtrdot \gamma$, $\beta \gtrdot \delta$, *and* $\alpha : \gamma = \beta : \delta$; *and conversely, if* $\alpha \gtrdot \gamma$, $\beta \gtrdot \delta$, $\alpha : \gamma = \beta : \delta$, *and* $\gamma = \delta$, *then* $\alpha = \beta$; *if* $\alpha \gtrdot \gamma$, $\beta \gtrdot \delta$, $\alpha : \gamma = \beta : \delta$, *and* $\alpha = \beta$, *then* $\gamma = \delta$, *provided* β *is not zero.*

For the direct theorem,

since $\alpha = \beta \gtrdot \gamma$, $\alpha \gtrdot \gamma$ and $\alpha : \gamma = \beta : \gamma$.

Since $\beta \gtrdot \gamma = \delta$, $\beta \gtrdot \delta$ and $\beta : \gamma = \beta : \delta$.

Therfor $\alpha : \gamma = \beta : \delta$.

For the first converse theorem,

since $\beta \gtrdot \delta = \gamma$, $\beta : \delta = \beta : \gamma$.

Hence, since $\alpha : \gamma = \beta : \delta$, $\alpha : \gamma = \beta : \gamma$.

Therfor $\alpha = \beta$.

Similarly the remaining theorem may be proved.

470. Theorem. *If* $\alpha \gtrdot \beta$, $\gamma = \delta$, *and* $\beta \gtrdot \gamma$, *then* $\alpha \gtrdot \gamma$, $\beta \gtrdot \delta$, $\alpha : \gamma \gtrdot \beta : \delta$, *and* $(\alpha : \gamma) : (\beta : \delta) = \alpha : \beta$; *and conversely, if* $\alpha \gtrdot \gamma$, $\beta \gtrdot \delta$, $\alpha : \gamma \gtrdot \beta : \delta$, *and* $\gamma = \delta$, *then* $\alpha \gtrdot \beta$.

For the direct theorem,

since $\alpha \gtrdot \beta \gtrdot \gamma$, $\alpha \gtrdot \gamma$, $\alpha : \gamma \gtrdot \beta : \gamma$,

and $(\alpha : \gamma) : (\beta : \gamma) = \alpha : \beta$. § 463

Since $\beta \gtrdot \gamma = \delta$, $\beta \gtrdot \delta$, and $\beta : \gamma = \beta : \delta$.

Therfor $\alpha : \gamma \gtrdot \beta : \delta$

and $(\alpha : \gamma) : (\beta : \gamma) = (\alpha : \gamma) : (\beta : \delta)$. §459.

Also $(\alpha : \gamma) : (\beta : \delta) = \alpha : \beta$.

For the converse theorem,

since $\beta \gtrdot \delta = \gamma$, $\beta : \delta = \beta : \gamma$.

Hence, since $\alpha : \gamma \gtrdot \beta : \delta$, $\alpha : \gamma \gtrdot \beta : \gamma$.

Therfor $\alpha \gtrdot \beta$.

471. Theorem. *If* $\alpha = \beta$, $\gamma \gtrdot \delta$, *and* $\beta \gtrdot \gamma$, *then* $\alpha \gtrdot \gamma$, $\beta \gtrdot \delta$, $\alpha : \gamma \lessdot \beta : \delta$, *and* $(\beta : \delta) : (\alpha : \gamma) = \gamma : \delta$, *provided* β *is not zero; and conversely, if* $\alpha \gtrdot \gamma$, $\beta \gtrdot \delta$, $\alpha : \gamma \lessdot \beta : \delta$, *and* $\alpha = \beta$, *then* $\gamma \gtrdot \delta$, *provided* β *is not zero.*

For the direct theorem,

since $\alpha = \beta \gtrdot \gamma$, $\alpha \gtrdot \gamma$, and $\alpha : \gamma = \beta : \gamma$.

Since $\beta \gtrdot \gamma \gtrdot \delta$, $\beta \gtrdot \delta$, $\beta : \delta \gtrdot \beta : \gamma$,

and $(\beta : \delta) : (\beta : \gamma) = \gamma : \delta$.

Therfor $\alpha : \gamma \lessdot \beta : \delta$

and $(\beta : \delta) : (\alpha : \gamma) = (\beta : \delta) : (\beta : \gamma)$.

Also $(\beta : \delta) : (\alpha : \gamma) = \gamma : \delta$.

For the converse theorem,

since $\beta = \alpha \gtrdot \gamma$, $\beta : \gamma = \alpha : \gamma$.

Hence, since $\alpha : \gamma \lessdot \beta : \delta$, $\beta : \gamma \lessdot \beta : \delta$.

Therfor $\gamma \gtrdot \delta$.

472. Theorem. *If* $\alpha \gtrdot \beta$, $\gamma \gtrdot \delta$, *and* $\beta \gtrdot \gamma$, *then* $\alpha \gtrdot \delta$, $\alpha : \delta \gtrdot \beta : \gamma$, *and*

$$(\alpha : \delta) : (\beta : \gamma) = (\alpha : \beta) \times (\gamma : \delta).$$

Since $\beta \gtrdot \gamma \gtrdot \delta$, $\beta \gtrdot \delta$, $\beta : \delta \gtrdot \beta : \gamma$, and $(\beta : \delta) : (\beta : \gamma) = \gamma : \delta$.

Since $\alpha \gtrdot \beta \gtrdot \delta$, $\alpha \gtrdot \delta$, $\alpha : \delta \gtrdot \beta : \delta$ and $(\alpha : \delta) : (\beta : \delta) = \alpha : \beta$.

Therfor $\alpha : \delta \mathrel{\dot{>}} \beta : \gamma$ and

$$(\alpha : \delta) : (\beta : \gamma) = [(\alpha : \delta) : (\beta : \delta)] \times [(\beta : \delta) : (\beta : \gamma)]$$
$$= (\alpha : \beta) \times (\gamma : \delta).$$

473. Theorem. *If* $\alpha \mathrel{\dot{>}} \beta$, $\gamma \mathrel{\dot{>}} \delta$, *and* $\alpha : \beta = \gamma : \delta$, *then* $\alpha\delta = \beta\gamma$; *and conversely, if* $\alpha \mathrel{\dot{>}} \beta$ *and* $\alpha\delta = \beta\gamma$, *then* $\gamma \mathrel{\dot{>}} \delta$ *and* $\alpha : \beta = \gamma : \delta$, *provided* δ *is not zero.*

For the direct theorem,

since $\alpha \mathrel{\dot{>}} \beta$, $\alpha\delta \mathrel{\dot{>}} \beta\delta$ and $(\alpha\delta) : (\beta\delta) = \alpha : \beta$.

Since $\gamma \mathrel{\dot{>}} \delta$, $\beta\gamma \mathrel{\dot{>}} \beta\delta$ and $(\beta\gamma) : (\beta\delta) = \gamma : \delta$.

Hence, since $\alpha : \beta = \gamma : \delta$, $(\alpha\delta) : (\beta\delta) = (\beta\gamma) : (\beta\delta)$.

Therfor $\alpha\delta = \beta\gamma$.

For the converse theorem, since $\alpha \mathrel{\dot{>}} \beta$,

$$\alpha\delta \mathrel{\dot{>}} \beta\delta \text{ and } (\alpha\delta) : (\beta\delta) = \alpha : \beta.$$

Hence, since $\alpha\delta = \beta\gamma$,

$$\beta\gamma \mathrel{\dot{>}} \beta\delta \text{ and } (\beta\gamma) : (\beta\delta) = (\alpha\delta) : (\beta\delta).$$

Hence $\gamma \mathrel{\dot{>}} \delta$ and $\gamma : \delta = (\beta\gamma) : (\beta\delta)$.

Therfor $\alpha : \beta = \gamma : \delta$.

474. Theorem. *If* $\alpha \mathrel{\dot{>}} \beta$, $\gamma \mathrel{\dot{>}} \delta$, *and* $\alpha : \beta \mathrel{\dot{>}} \gamma : \delta$, *then* $\alpha\delta \mathrel{\dot{>}} \beta\gamma$ *and* $(\alpha\delta) : (\beta\gamma) = (\alpha : \beta) : (\gamma : \delta)$; *and conversely, if* $\gamma \mathrel{\dot{>}} \delta$ *and* $\alpha\delta \mathrel{\dot{>}} \beta\gamma$, *then* $\alpha \mathrel{\dot{>}} \beta$ *and* $\alpha : \beta \mathrel{\dot{>}} \gamma : \delta$.

For the direct theorem,

since $\alpha \mathrel{\dot{>}} \beta$, $\alpha\delta \mathrel{\dot{>}} \beta\delta$ and $(\alpha\delta) : (\beta\delta) = \alpha : \beta$.

Since $\gamma \mathrel{\dot{>}} \delta$, $\beta\gamma \mathrel{\dot{>}} \beta\delta$ and $(\beta\gamma) : (\beta\delta) = \gamma : \delta$.

Hence, since $\alpha : \beta \mathrel{\dot{>}} \gamma : \delta$, $(\alpha\delta) : (\beta\delta) \mathrel{\dot{>}} (\beta\gamma) : (\beta\delta)$ and $[(\alpha\delta) : (\beta\delta)] : [(\beta\gamma) : (\beta\delta)] = (\alpha : \beta) : (\gamma : \delta)$. § 469.

Therfor $\alpha\delta \mathrel{\dot{>}} \beta\gamma$ and $(\alpha\delta) : (\beta\gamma) = (\alpha : \beta) : (\gamma : \delta)$.

For the converse theorem,

since $\gamma \mathrel{\dot{>}} \delta$, $\beta\gamma \mathrel{\dot{>}} \beta\delta$ and $(\beta\gamma) : (\beta\delta) = \gamma : \delta$.

Hence, since $\alpha\delta \gtrdot \beta\gamma$,

$$\alpha\delta \gtrdot \beta\delta \text{ and } (\alpha\delta) : (\beta\delta) \gtrdot (\beta\gamma) : (\beta\delta).$$

Therfor $\alpha \gtrdot \beta$ and $\alpha : \beta = (\alpha\delta) : (\beta\delta)$.

Therfor $\alpha : \beta \gtrdot \gamma : \delta$.

475. By comparing §§ 409–474 with §§ 75–128, the analogy may be seen that exists between the theory of products and quotients on the one hand and that of sums and differences on the other. Also certain points of divergence between the two theories will appear.

476. Theorem. *If* $\alpha \gtrdot \beta$, *and* $\alpha : \beta \gtrdot \gamma$, *then* $\alpha \gtrdot \beta\gamma$ *and* $\alpha : (\beta\gamma) = (\alpha : \beta) : \gamma$.

This follows from § 474 by supposing $\delta = 1$.

477. Theorem. *If* $\alpha \gtrdot \beta$ *and* $\gamma \gtrdot \alpha : \beta$, *then* $\gamma\beta \gtrdot \alpha$ *and* $(\gamma\beta) : \alpha = \gamma : (\alpha : \beta)$.

478. Theorem. *If* $\alpha \gtrdot \gamma$ *and* $\beta \gtrdot \gamma$, *then* $\alpha + \beta \gtrdot \gamma$ *and* $(\alpha + \beta) : \gamma = (\alpha : \gamma) + (\beta : \gamma)$; *and conversely, if* $\alpha \gtrdot \gamma$ *and* $\alpha + \beta \gtrdot \gamma$, *then* $\beta \gtrdot \gamma$ *and*

$$(\alpha + \beta) : \gamma = (\alpha : \gamma) + (\beta : \gamma).$$

For the direct theorem,

since $\alpha \gtrdot \gamma$, $\alpha = \varphi\gamma$ and $\alpha : \gamma = \varphi$.

Since $\beta \gtrdot \gamma$, $\beta = \chi\gamma$ and $\beta : \gamma = \chi$.

Hence $\alpha + \beta = \varphi\gamma + \chi\gamma = (\varphi + \chi)\gamma$.

Therfor $\alpha + \beta \gtrdot \gamma$ and $(\alpha + \beta) : \gamma = \varphi + \chi$.

Or $(\alpha + \beta) : \gamma = (\alpha : \gamma) + (\beta : \gamma)$.

For the converse theorem,

since $\alpha \gtrdot \gamma$, $\alpha = \varphi\gamma$.

Since $\alpha + \beta \gtrdot \gamma$, $\alpha + \beta = \psi\gamma$.

Hence $(\alpha + \beta) - \alpha = \psi\gamma - \varphi\gamma = (\psi - \varphi)\gamma$.

Or $\beta = (\psi - \varphi)\gamma$.

Therfor $\beta \mathrel{\dot{>}} \gamma$ and by the direct theorem

$$(\alpha + \beta) : \gamma = (\alpha : \gamma) + (\beta : \gamma).$$

This theorem is the **Distributiv Law for Division and Addition.**

479. Theorem. *If $\alpha \mathrel{\dot{>}} \gamma$ and $\beta \mathrel{\dot{>}} \gamma$, then $\alpha - \beta \mathrel{\dot{>}} \gamma$ and $(\alpha - \beta) : \gamma = (\alpha : \gamma) - (\beta : \gamma)$; and conversely, if $\alpha \mathrel{\dot{>}} \gamma$ and $\alpha - \beta \mathrel{\dot{>}} \gamma$, then $\beta \mathrel{\dot{>}} \gamma$ and $(\alpha - \beta) : \gamma = (\alpha : \gamma) - (\beta : \gamma)$; if $\beta \mathrel{\dot{>}} \gamma$ and $\alpha - \beta \mathrel{\dot{>}} \gamma$, then $\alpha \mathrel{\dot{>}} \gamma$ and $(\alpha - \beta) : \gamma = (\alpha : \gamma) - (\beta : \gamma)$.*

480. Theorem. *If $\alpha \mathrel{\dot{>}} \gamma$, $\beta \mathrel{\dot{>}} \gamma$, and λ and μ ar any two integers, then $\lambda\alpha + \mu\beta \mathrel{\dot{>}} \gamma$ and $(\lambda\alpha + \mu\beta) : \gamma = \lambda(\alpha : \gamma) + \mu(\beta : \gamma)$* (1).

481. Theorem. *If $\alpha \mathrel{\dot{>}} \gamma$, $\beta \mathrel{\dot{>}} \gamma$, and λ and μ ar any two integers, then $\lambda\alpha - \mu\beta \mathrel{\dot{>}} \gamma$ and $(\lambda\alpha - \mu\beta) : \gamma = \lambda(\alpha : \gamma) - \mu(\beta : \gamma)$.*

482. Theorem. *If $\alpha \mathrel{\dot{>}} \beta$ and $\gamma \mathrel{\dot{>}} \delta$, then $\alpha\delta + \beta\gamma \mathrel{\dot{>}} \beta\delta$ and $(\alpha\delta + \beta\gamma) : (\beta\delta) = (\alpha : \beta) + (\gamma : \delta)$.*

Since $\alpha \mathrel{\dot{>}} \beta$, $\alpha\delta \mathrel{\dot{>}} \beta\delta$ and $(\alpha\delta) : (\beta\delta) = \alpha : \beta$.

Since $\gamma \mathrel{\dot{>}} \delta$, $\beta\gamma \mathrel{\dot{>}} \beta\delta$ and $(\beta\gamma) : (\beta\delta) = \gamma : \delta$.

Therfor $\alpha\delta + \beta\gamma \mathrel{\dot{>}} \beta\delta$

and $(\alpha\delta + \beta\gamma) : (\beta\delta) = (\alpha\delta) : (\beta\delta) + (\beta\gamma) : (\beta\delta)$.

Also $(\alpha\delta + \beta\gamma) : (\beta\delta) = (\alpha : \beta) + (\gamma : \delta)$.

483. Theorem. *If $\alpha \mathrel{\dot{>}} \beta$ and $\gamma \mathrel{\dot{>}} \delta$, then $\alpha\delta - \beta\gamma \mathrel{\dot{>}} \beta\delta$ and $(\alpha\delta - \beta\gamma) : (\beta\delta) = (\alpha : \beta) - (\gamma : \delta)$.*

484. Theorem. *If $\alpha \mathrel{\dot{>}} \beta$ and α and β ar both positiv or both negativ, then $\alpha : \beta$ is positiv; if either α or β is*

---

(1) See Legendre, p. 3; Chrystal, p. 39.

positiv and the other negativ, then $\alpha : \beta$ *is negativ; and conversely, if the quotient* $\alpha : \beta$ *is positiv,* α *and* β *ar either both positiv or both negativ; if* $\alpha : \beta$ *is negativ, either* α *or* β *is positiv and the other negativ.*

Suppose $\alpha \gtrdot \beta$ and that α and β ar both positiv.

Then $\alpha = \varphi\beta$ and $\alpha : \beta = \varphi$.

Now φ is either positiv, zero, or negativ.

But φ cannot be zero, since then $\varphi\beta$, and hence α, would be zero.

Similarly φ cannot be negativ.

Therfor φ is positiv.

The other parts of the direct theorem ar proved in a similar manner.

The converse theorem is proved by the method of exhaustion.

485. Theorem. *If* $\alpha \gtrdot \beta$, *then* $-\alpha \gtrdot \beta$ *and* $(-\alpha) : \beta = -(\alpha : \beta)$.

Since $\alpha \gtrdot \beta$, $\alpha = \varphi\beta$ and $\alpha : \beta = \varphi$.

Hence $\quad -\alpha = -(\varphi\beta) = (-\varphi)\beta$.

Therfor $-\alpha \gtrdot \beta$ and $(-\alpha) : \beta = -\varphi = -(\alpha : \beta)$.

486. Since the expressions $(-\alpha) : \beta$ and $-(\alpha : \beta)$ ar equal, the parentheses may be omitted and both be written $-\alpha : \beta$.

487. Theorem. *If* $\alpha \gtrdot \beta$, *then* $\alpha \gtrdot -\beta$ *and* $\alpha : (-\beta) = -(\alpha : \beta)$.

488. Theorem. *If* $\alpha \gtrdot \beta$, *then* $-\alpha \gtrdot -\beta$ *and* $(-\alpha) : (-\beta) = \alpha : \beta$.

489. Theorem. $\alpha \gtrdot -1$ *and* $\alpha : (-1) = -\alpha$.

490. From the theorems of §§ 170, 485, 487, 488 we derive the following equalities:

$(+\alpha):(+\beta)=+\alpha:\beta,$
$(-\alpha):(+\beta)=-\alpha:\beta,$
$(+\alpha):(-\beta)=-\alpha:\beta,$
$(-\alpha):(-\beta)=+\alpha:\beta,$ provided $\alpha \dot{>} \beta.$

These equalities constitute the **Rule of Signs for Division,** which is usually briefly stated: *In division like signs giv* $+$, *unlike signs* $-$.

In particular we hav $(-a):b=a:(-b)=-(a:b)$ and $(-a):(-b)=a:b$, provided $a \dot{>} b$.

491. Theorem. *If* $\alpha>\beta$ *and* γ *is positiv, then* $\alpha:\gamma>\beta:\gamma$, *provided* α *and* β *ar both divisible by* γ.

Since $\alpha \dot{>} \gamma$, $\alpha=\varphi\gamma$ and $\alpha:\gamma=\varphi$.

Since $\beta \dot{>} \gamma$, $\beta=\chi\gamma$ and $\beta:\gamma=\chi$.

Since $\alpha>\beta$, $\varphi\gamma>\chi\gamma$.

Therfor, since γ is positiv, $\varphi>\chi$. § 345.

Or $\alpha:\gamma>\beta:\gamma$.

492. Theorem. *If* $\alpha>\beta$ *and* γ *is negativ, then* $\alpha:\gamma<\beta:\gamma$, *provided* α *and* β *ar both divisible by* γ.

493. Theorem. *If* $\alpha>\beta$, *and* β *and* γ *ar positiv, then* $\gamma:\alpha<\gamma:\beta$, *provided* γ *is divisible by both* α *and* β.

Since $\gamma \dot{>} \alpha$, $\gamma=\varphi\alpha$ and $\gamma:\alpha=\varphi$.

Since $\gamma \dot{>} \beta$, $\gamma=\chi\beta$ and $\gamma:\beta=\chi$.

Hence $\varphi\alpha=\chi\beta$. (1)

Since β is positiv, α is positiv. § 201.

Since β and γ ar positiv, χ is positiv.

Hence, from (1), since $\alpha>\beta$ and α and χ ar positiv, $\varphi<\chi$. § 353.

Therfor $\gamma : \alpha < \gamma : \beta$.

494. Theorem. *If* $\alpha > \beta$ *and* $\gamma = \delta$; *then, if* γ *is positiv,* $\alpha : \gamma > \beta : \delta$; *if* γ *is negativ,* $\alpha : \gamma < \beta : \delta$.

This follows from §§ 491, 459, 492.

495. Theorem. *If* $\alpha = \beta$, $\gamma > \delta$, *and* α *and* δ *ar positiv, then* $\alpha : \gamma < \beta : \delta$.

496. Theorem. *If* $\alpha \dot{>} \beta$, $|\alpha| \dot{>} |\beta|$ *and* $|\alpha| : |\beta| = |\alpha : \beta|$.

Since $\alpha \dot{>} \beta$, $\alpha = \varphi\beta$ and $\alpha : \beta = \varphi$.

Hence $|\alpha| = |\varphi\beta| = |\varphi|\,|\beta|$.

Therfor $|\alpha| \dot{>} |\beta|$ and $|\alpha| : |\beta| = |\varphi| = |\alpha : \beta|$.

497. Theorem. *If an integer is not zero, each of its factors, except the integer itself and its opposit, is numerically less than the integer.*

Let $\alpha \dot{>} \beta$, wher $\alpha \neq 0$ and $|\alpha| \neq |\beta|$.

Then $\alpha = \varphi\beta$ and $|\alpha| = |\varphi\beta| = |\varphi|\,|\beta|$.

Now either $|\varphi| = 0$, $|\varphi| = 1$, or $|\varphi| > 1$.

If $|\varphi| = 0$, $\varphi = 0$ and $\alpha = \varphi\beta = 0$. § 398.

But $\alpha \neq 0$. Hence $|\varphi| \neq 0$.

If $|\varphi| = 1$, $|\alpha| = |\beta|$.

Hence in this case β is either α or its opposit. § 399.

Ther remains only the case when $|\varphi| > 1$.

In this case $|\varphi\beta| > |\beta|$. § 406.

That is, $|\alpha| > |\beta|$, or $|\beta| < |\alpha|$.

Hence the theorem is proved.

498. Theorem. *No integer except zero is divisible by an integer numerically greater than itself.*

499. For example, the symbols $(-3:9)$ $(-4):(-8)$, $5:(-7)$ ar, as yet, meaningless.

500. Theorem. *Unity has no positiv factor except unity.*

501. Theorem. *If* $\nu = \alpha\beta$ *and* $1 < |\alpha| < |\nu|$, *then* $1 < |\beta| < |\nu|$.

For, since $|\nu| > 1$, $|\nu| > 0$. (1)

Now $|\nu| = |\alpha\beta| = |\alpha|\,|\beta|$. (2)

If $|\beta| = 0$, $|\nu| = 0$, which contradicts (1).

If $|\beta| = 1$, $|\nu| = |\alpha|$, which contradicts the hypothesis.

If $|\beta| = |\nu|$, $|\nu| = |\alpha|\,|\nu|$ and hence $|\alpha| = 1$, which contradicts the hypothesis.

If $|\beta| > |\nu|$, then, since $|\alpha\beta| > |\beta|$ (§ 406). $|\alpha\beta| > |\nu|$, which contradicts (2).

Therfor $1 < |\beta| < |\nu|$.

502. Theorem. *If* $\alpha \gtrdot \beta$ *and* $\beta \gtrdot \alpha$, $|\alpha| = |\beta|$.

By the hypothesis neither α nor β is zero.

Hence, since $\alpha \gtrdot \beta$, $|\alpha| \geqq |\beta|$. § 497.

Also, since $\beta \gtrdot \alpha$, $|\alpha| \leqq |\beta|$.

Therfor $|\alpha| = |\beta|$.

503. Theorem. *If* α *and* β *ar any two integers, except that* β *may not be zero, an infinit number of pairs of integers* λ, μ *can be found such that* $\lambda\beta < \alpha < \mu\beta$.

Consider the multiples of β

$$\cdots, (-3)\beta, (-2)\beta, (-1)\beta, 0\beta, 1\beta, 2\beta, 3\beta, \cdots$$

These multiples increase from left to right, if β is positiv, and from right to left, if β is negativ. Among them ar found the products $\alpha\beta$ and $(-\alpha)\beta$.

Ther ar three cases to be considerd.

Case 1. $\alpha = 0$. Then $-|\beta| < \alpha < |\beta|$.

This givs, if β is positiv, $(-1)\beta < \alpha < 1 \times \beta$; if β is negativ, $1 \times \beta < \alpha < (-1)\beta$.

Case 2. $|\beta| = 1$. Since $\alpha - 1 < \alpha < \alpha + 1$, we hav $(\alpha - 1)|\beta| < \alpha < (\alpha + 1)|\beta|$.

This givs, if β is positiv, $(\alpha - 1)\beta < \alpha < (\alpha + 1)\beta$; if β is negativ, $(-(\alpha - 1))\beta < \alpha < (-(\alpha + 1))\beta$.

Case 3. $\alpha \neq 0$ and $|\beta| > 1$. Here

$$0 < |\alpha| < |\alpha||\beta| \text{(1)}.$$

This case may be divided into four subcases.

1°. α positiv, β positiv. The above result becomes $0 < \alpha < \alpha\beta$ or $0 \cdot \beta < \alpha < \alpha\beta$.

2°. α positiv, β negativ. Here $0 \cdot \beta < \alpha < (-\alpha)\beta$.

3°. α negativ, β positiv. Here $0 < -\alpha < (-\alpha)\beta$, which givs $0 > \alpha > \alpha\beta$, or $\alpha\beta < \alpha < 0 \cdot \beta$.

4°. α negativ, β negativ. Here

$$(-\alpha)\beta < \alpha < 0 \cdot \beta.$$

In each case, then, we hav found two multiples of β between which α lies. The coefficients of these multiples ar a pair of numbers λ, μ, such that

$$\lambda\beta < \alpha < \mu\beta.$$

Again, if β is positiv and κ is any integer less than λ and ν any integer greater than μ, $\kappa\beta < \lambda\beta$ and $\mu\beta < \nu\beta$. Hence $\kappa\beta < \alpha < \nu\beta$.

Similarly, if β is negativ.

Therfor an infinit number of such pairs of integers can be found. §§ 282, 137.

---

(1) See Stolz und Gmeiner, p. 25.

504. When α and β ar positiv, μ is positiv. The preceding theorem then givs as a particular result the following:

Theorem. *If a and b ar two primary numbers and $a > b$, a primary number n can be found such that $nb > a$.*

This theorem is sometimes calld the **Axiom of Archimedes** ([1]).

505. The theorem of § 503 furnishes us a means for determining whether $\alpha \gtrdot \beta$ and, if $\alpha \gtrdot \beta$, for finding the quotient $\alpha : \beta$.

If α is a multiple of β, it must be one of the multiples that lie between two particular multiples $\lambda\beta$ and $\mu\beta$, which ar easily found. We hav then to consider only the limited number of multiples that lie between $\lambda\beta$ and $\mu\beta$. If one of these is equal to α, $\alpha \gtrdot \beta$ and $\alpha : \beta$ is the coefficient of that multiple. If none of these is equal to α, $\alpha \not\gtrdot \beta$.

Thus, to find whether 7 is divisible by 2, we need consider only the products that lie between 0×2 and 7×2. We have $1 \times 2 = 2$, $2 \times 2 = 4$, $3 \times 2 = 6$, $4 \times 2 = 8$. Now, if $n > 4$, $n \times 2 > 4 \times 2$. Hence all higher products ar greater than 8, and therfor greater than 7.

Since, then, ther is no multiple of 2 which equals 7, $7 \not\gtrdot 2$.

([1]) See D. Hilbert, "Grundlagen der Geometrie," 1899, English translation, "The Foundations of Geometry," 1902, by E. J. Townsend, p. 25.

The symbol $7:2$ is, then, at present meaningless. Similarly the symbols $7:(-5)$, $4:3$, $(-5):(-4)$ ar meaningless.

506. Definition. Division. The process described in § 505 for finding whether α is divisible by β, and, if divisible, for finding the quotient, is calld **division.**

Thus the quotient $\alpha:\beta$, when it exists, is obtaind from α and β by an operation, the operation of dividing α by β. In this operation α is the operand, the passiv element, or **dividend**, and β the operator, the activ element, or **divisor.**

When α and β ar the primary numbers a and b, their quotient $a:b$, when it exists, may be obtaind by dividing a group of a objects into b groups, the numbers of objects in the latter groups being the same, $a:b$.

Or we may divide a group of a objects into groups of b each, the number of these groups being $a:b$ (1).

These results follow directly from § 455.

507. By § 490 the quotient of a positiv and a negativ integer, or of a negativ and a positiv integer, or of two negativ integers, may be found when the quotient of the corresponding positiv integers is known, if the latter quotient exists.

508. Division Table. In order to make a table of the existing quotients of all possible pairs of the numbers 1, 2, 3, ⋯, 10, we divide a square into compartments as in § 74 and write the numbers 1, 2, 3, ⋯, 10

(1) See Tannery, p. 17.

along the top and left-hand sides of the square as in that article.

In each compartment we wish to place the number which is the quotient of the two numbers found respectivly at the left of the row and at the top of the column in which the compartment lies, if this quotient exists.

Thus in the eighth row of compartments we should write the numbers that represent the quotients $8:1$, $8:2$, $8:3$, $8:4$, $\cdots$

To find these quotients we may use the multiplication table to advantage. We notice that in the compartments of that table the 8's lie in the first, second, fourth, and eighth columns. The 8 of the first column is in the eighth row; hence $8:1=8$. The 8 of the second column is in the fourth row; hence $8:2=4$. Ther is no 8 in the third column; hence ther is no quotient $8:3$. The 8 of the fourth column is in the second row; hence $8:4=2$. And so on.

Hence the eighth row of our new table will contain the numbers 8, 4, 2, 1 in the first, second, fourth, and eighth columns, and none in the others.

In a similar manner we may form the rest of the table, given on the next page.

509. Theorem. *The operation of division is inverse to multiplication* (1).

This follows immediately from § 453.

(1) See Schubert, "Encyklopädie," p. 16.

DIVISION TABLE.

	1	2	3	4	5	6	7	8	9	ϒ	0	Ꞓ	ƨ	Ɏ	Ͻ	10
1	1															
2	2	1														
3	3		1													
4	4	2		1												
5	5				1											
6	6	3	2			1										
7	7						1									
8	8	4		2				1								
9	9		3						1							
ϒ	ϒ	5			2					1						
0	0										1					
Ꞓ	Ꞓ	6	4	3		2						1				
ƨ	ƨ												1			
Ɏ	Ɏ	7					2							1		
Ͻ	Ͻ		5		3										1	
10	10	8		4				2								1

510. Theorem. *Multiplication is inverse to division.*

511. Definition. The operation of multiplication, which was defined (§ 297) as a repeated performance of the direct operation addition is also calld a **direct,** or **synthetic,** operation.

On the other hand, in division, having given two integers, of which the second is a factor of the first, the object is to find a third integer which multiplied by the second makes the first. Every question in division is, then, turnd into a question of multiplication. For this reason division is calld an **indirect,** or **analytic,** operation.

For reasons the same as above given $\alpha \times \beta$ and $\alpha : \beta$ ar calld respectivly **synthetic** and **analytic combinations** of the integers α and β.

The opposition in nature between multiplication and division may be shown by the equality $\gamma = \alpha\beta$.

If α and β ar given to find γ, we get the result by multiplication.

If γ and β ar given to find α, or γ and α given to find β, we get the result by division, the operations in these two cases being of the same nature because of the commutativ law for multiplication.

To distinguish multiplication and division on the one hand from addition and subtraction on the other, the former ar calld the operations of the **second step** (1), or **second degree,** the latter the operations of the **first step,** or **first degree.**

The operation of multiplication is always possible. The operation of division is possible only when the passiv number is a multiple of the activ number.

512. Theorem. *When the operation of division shows that* $\alpha \not\mathrel{:\!>} \beta$, *a pair of integers* φ, χ *can be found, and only one pair, such that* $\varphi\beta$ *is the greatest of all the multiples of* β *which ar less* (2) *than* α, $\chi\beta$ *the least of*

(1) See Schubert, "Encyklopädie," p. 14; Stolz und Gmeiner, p. 9.

(2) See Chrystal, p. 42.

all the multiples of β *which ar greater than* α, *and* $\chi - \varphi = \pm 1$ (1), *according as* β *is positiv or negativ.*

We hav $\lambda\beta < \alpha < \mu\beta$. § 503.

Of course all those multiples of β which ar less than $\lambda\beta$ ar also less than α and those greater than $\mu\beta$ ar greater than α.

Now of all the multiples which ar less than α, from $\lambda\beta$ up, pick out the greatest and let its coefficient be φ. Then $\varphi\beta$ is the greatest of all the multiples of β that ar less than α.

Also, of all the multiples which ar greater than α, from $\mu\beta$ down, pick out the smallest and let its coefficient be χ. Then $\chi\beta$ is the smallest of all the multiples of β that ar greater than α.

We hav, then, $\varphi\beta < \alpha < \chi\beta$, and hence $\varphi\beta < \chi\beta$.

Now $\varphi\beta$ and $\chi\beta$ ar consecutiv multiples of β.

For, suppose ther existed an intermediate multiple, $\psi\beta$, so that $\varphi\beta < \psi\beta < \chi\beta$.

Then either $\psi\beta < \alpha$ or $\psi\beta > \alpha$.

If $\psi\beta < \alpha$, then $\varphi\beta$ would not be the greatest of all the multiples of β which ar less than α.

If $\psi\beta > \alpha$, then $\chi\beta$ would not be the smallest of all the multiples of β which are greater than α.

Hence ther is no multiple of β between $\varphi\beta$ and $\chi\beta$ and these two multiples ar consecutiv.

Therfor, if β is positiv, $\chi = \varphi + 1$ and $\chi - \varphi = 1$;
if β is negativ, $\chi = \varphi - 1$ and $\chi - \varphi = -1$.

Only one such pair of integers can be found, since

(1) See Stolz und Gmeiner, p. 24.

ther is only one greatest, and one least, among a set of integers, no two of which ar equal.

513. Definition. Besides the process previously calld division (§ 506), the process for finding the integers φ and χ, when $\alpha \not\gtrdot \beta$, as described in the last article, is also calld **division**; α is the **dividend**, β the **divisor**, φ the **lower quotient**, χ the **upper quotient**.

To distinguish division in the sense of § 506 from division in the present sense, the former may be calld **exact division** and the latter **approximate division** (1). In exact division the dividend is said to be **exactly divisible** by the divisor.

514. Definition. When α is divided by β giving a quotient φ, either exact quotient, lower approximate quotient, or upper approximate quotient, the difference $\alpha - \varphi\beta$ is calld the **remainder** (2).

515. Theorem. *When α is divided by β giving a quotient φ, either exact, lower approximate, or upper approximate, if ρ is the corresponding remainder, then* $\alpha = \varphi\beta + \rho$.

516. Theorem. *In exact division the remainder is zero; in approximate division the remainder corresponding to the lower quotient is positiv, the remainder corresponding to the upper quotient is negativ.*

For in the first case $\alpha = \varphi\beta$, in the second $\alpha > \varphi\beta$, in the third $\alpha < \varphi\beta$.

517. When in approximate division the words

(1) See Stolz und Gmeiner, p. 25.

(2) See Chrystal, p. 42; Tannery, p. 80; Stolz und Gmeiner, p. 25.

"quotient" and "remainder" ar used without qualification, the lower quotient and positiv remainder ar usually meant.

518. Theorem. *When α is approximately divided by β, the sum of the numerical values of the positiv and negativ remainders is equal to the numerical value of β.*

Let ρ and σ be the positiv and negativ remainders respectivly.

Then $|\rho| + |\sigma| = \rho + \bar{\sigma} = \rho - \sigma$.

But $\alpha = \varphi\beta + \rho$ and $\alpha = \chi\beta + \sigma$.

Hence $\varphi\beta + \rho = \chi\beta + \sigma$.

Therfor $\rho - \sigma = \chi\beta - \varphi\beta = (\chi - \varphi)\beta$
$= \pm 1 \cdot \beta$, according as β is positiv or negativ,
$= \pm \beta$.

Hence $|\rho| + |\sigma| = |\beta|$.

519. Theorem. *When α is approximately divided by β, both remainders ar numerically less than β.*

This follows from §§ 518, 264, 234.

520. Theorem. *If, when two integers ar approximately divided by the same integer, the positiv remainders ar equal, then the negativ remainders ar also equal, and conversely.*

521. Theorem. *If $\alpha = \varphi\beta + \rho$, wher $\beta \neq 0$ and $|\rho| < |\beta|$; then, if ρ is zero, φ is the exact quotient of α divided by β; if ρ is positiv, φ is the lower approximate quotient; and, if ρ is negativ, φ is the upper approximate quotient; in each case ρ is the corresponding remainder.*

First, if $\rho = 0$, $\alpha = \varphi\beta + \rho = \varphi\beta$.

Therfor $\varphi = \alpha : \beta$.

Second, suppose ρ is positiv.

Then $\varphi\beta < \alpha$.

Moreover $\varphi\beta$ is the greatest of all the multiples of β which ar less than α.

For $(\varphi \pm 1)\beta > \varphi\beta$, the upper or the lower sign being used according as β is positiv or negativ.

§§ 332, 336.

Also, since $\pm\beta > \rho$, $\varphi\beta \pm \beta > \varphi\beta + \rho$.

Or $(\varphi \pm 1)\beta > \alpha$.

Hence the next higher multiple of β after $\varphi\beta$ is greater than α.

Therfor $\varphi\beta$ is the greatest of all the multiples of β which ar less than α, and φ is the lower approximate quotient of α divided by β.

Similarly, if ρ is negativ, it may be shown that φ is the upper approximate quotient.

In each case we have $\alpha - \varphi\beta = \rho$ and therfor ρ is the corresponding remainder.

522. Theorem. *If α and β ar any two integers, except that β may not be zero; then α can be written in the form $\varphi\beta + \rho$, wher $|\rho| < |\beta|$, in one, and only one, way, when α is exactly divisible by β; and in two ways, and no more, when α is not exactly divisible by β, one with ρ positiv and one with ρ negativ.*

It follows from § 515 that one such way exists when α is exactly divisible by β, φ being the quotient $\alpha : \beta$ and ρ being zero, and two ways when α is not exactly divisible by β, in one way φ being the lower quotient and ρ the corresponding positiv remainder, in the other φ being the upper quotient and ρ the corresponding negativ remainder.

To prove that no other such form of expressing α can be obtaind when α is exactly divisible by β, suppose $\psi\beta + \rho$ is any such form.

We hav also $\alpha = \varphi\beta$.

Hence $\psi\beta + \rho = \varphi\beta$.

Therfor $\rho = \varphi\beta - \psi\beta = (\varphi - \psi)\beta$.

Hence $\rho \not< \beta$. But $|\rho| < |\beta|$. Therfor $\rho = 0$. § 498.

Hence ψ is the exact quotient of α divided by β. § 521.

This form is, then, the same as the other.

Next suppose α is not exactly divisible by β and let $\varphi\beta + \rho$ be any form of expressing α.

Then $\rho \neq 0$. For, if $\rho = 0$, α would be exactly divisible by β.

Hence ρ is either positiv or negativ.

If ρ is positiv, φ is the lower approximate quotient and ρ the corresponding remainder. § 521.

If ρ is negativ, φ is the upper approximate quotient and ρ the corresponding remainder.

Our theorem is, then, proved.

523. Theorem. *If α is exprest in the form $\varphi\beta + \rho$, wher $|\rho| < |\beta|$, then α is or is not exactly divisible by β, according as ρ is or is not zero.*

524. Theorem. *If α be approximately divided by β,* *giving a quotient* φ *and remainder* ρ, *then, when* $\alpha\theta$ *is divided by* $\beta\theta$, *a quotient, lower or upper, will be* φ(1) *and the corresponding remainder* $\rho\theta$.

For we hav $\alpha = \varphi\beta + \rho$, wher $|\rho| < |\beta|$.

Hence $\alpha\theta = (\varphi\beta)\theta + \rho\theta$.

Or $\alpha\theta = \varphi(\beta\theta) + \rho\theta$.

Also, since $|\theta|$ is positiv, $|\rho||\theta| < |\beta||\theta|$.

That is, $|\rho\theta| < |\beta\theta|$.

Therfor, when $\alpha\theta$ is divided by $\beta\theta$, a quotient, lower or upper, must be φ and the corresponding remainder $\rho\theta$. § 521.

525. The most important particular case of the preceding investigation is when α and β ar both positiv integers, or primary numbers.

Suppose a and b ar any two primary numbers and $a > b$.

If a is exactly divisible by b, $a = qb$.

If a is not exactly divisible by b, let q be the lower quotient, t the upper, r the positiv remainder, and $-s$ the negativ remainder.

Then $a = qb + r$, $a = tb - s$, $r + s = b$, and r and s ar both less than b.

We hav also the following additional theorems for primary numbers, or positiv integers.

526. Theorem. *If* α *and* β *ar positiv,* $\alpha > \beta$, *and* $\alpha \not\gtrdot \beta$, *then both approximate quotients obtaind by dividing* α *by* β *ar positiv.*

---

(1) See Stolz und Gmeiner, p. 34.

Let φ be the lower approximate quotient and ρ the corresponding remainder.

Then $\alpha = \varphi\beta + \rho$, wher $\alpha > \beta > \rho$.

Hence $\varphi\beta = \alpha - \rho$, wher $\alpha > \rho$.

Hence $\varphi = (\alpha - \rho) : \beta$.

Since $\alpha > \rho$, $\alpha - \rho$ is positiv.

Since, therfor, φ is the quotient of two positiv integers, φ is positiv.

Similarly we may prove that the upper quotient is positiv.

527. Theorem. *If α and β ar positiv, $\alpha > \beta$, and $\alpha \not\vdots \beta$, and the lower approximate quotient φ obtaind when α is divided by β is greater than the corresponding remainder, then, when α is divided by φ, the lower approximate quotient is β and the remainder the same as when α is divided by β.*

Let ρ be the positiv remainder obtaind when α is divided by β.

Then $\alpha = \varphi\beta + \rho$, wher $\rho < \varphi$.

Hence $\alpha = \beta\varphi + \rho$.

The theorem then follows from § 521.

528. Theorem. *If α and β ar positiv, $\alpha > \beta$, and $\alpha \not\vdots \beta$, and the lower approximate quotient φ obtaind when α is divided by β is greater than β, then, when α is divided by φ, the lower approximate quotient is β and the remainder the same as when α is divided by β.*

For, if $\varphi > \beta$ and $\beta > \rho$, $\quad \varphi > \rho$.

529. Theorem. *If α and β ar positiv, $\alpha > \beta$, and $\alpha \not\vdots \beta$, and the upper approximate quotient χ obtaind* *when α is divided by β is greater than the corresponding remainder, then, when α is divided by χ, the upper approximate quotient is β and the remainder the same as when α is divided bv β.*

530. Theorem. *If α and β ar positiv, $\alpha > \beta$, and $\alpha \not\gtrdot \beta$, and the upper approximate quotient χ obtaind when α is divided by β is greater than β, then, when α is divided by χ, the upper approximate quotient is β and the remainder the same as when α is divided by β.*

531. Theorem. *If $1 < \beta < \alpha$ and φ is the exact or lower approximate quotient obtaind by dividing α by β, then both φ and $\varphi + 1$ ar less than α and $(\varphi + 1)\beta > \alpha$.*

Since $\beta > 1$, $\beta \geqq 2$.

Hence $\varphi\beta \geqq \varphi \times 2$. §§ 330, 335.

Or $\varphi\beta \geqq \varphi + \varphi$.

Now $\alpha \geqq \varphi\beta$.

If $\alpha = \varphi\beta$, $\varphi > 1$. § 501.

Hence $\varphi + \varphi > \varphi + 1$. § 216.

Hence in this case we hav

$$\alpha = \varphi\beta \geqq \varphi + \varphi > \varphi + 1.$$

Therfor $\alpha > \varphi + 1$.

Moreover, since $\varphi + 1 > \varphi$, $(\varphi + 1)\beta > \varphi\beta$. § 332.

Therfor $(\varphi + 1)\beta > \alpha$.

If $\alpha > \varphi\beta$, $\varphi \geqq 1$.

Hence $\varphi + \varphi \geqq \varphi + 1$.

Hence in this case $\alpha > \varphi\beta \geqq \varphi + \varphi \geqq \varphi + 1$.

Therfor $\alpha > \varphi + 1$.

Moreover, since $\varphi + 1$ is the upper approximate

quotient obtaind by dividing α by β, $(\varphi + 1)\beta > \alpha$.
§ 512.

In both cases, since $\varphi < \varphi + 1 < \alpha$, $\varphi < \alpha$.

532. Theorem. *If $\alpha > 2$ and β is the integer one greater than the exact or lower approximate quotient obtaind by dividing α by 2, then $1 < \beta < \alpha$ and $\beta\beta > \alpha$.*

Let the exact or lower approximate quotient obtaind by dividing α by 2 be φ.

Then $\varphi + 1 < \alpha$ and $(\varphi + 1)2 > \alpha$. § 531.

Also $\varphi + 1 > 1$.

Hence, if $\beta = \varphi + 1$, $1 < \beta < \alpha$ and $\beta \times 2 > \alpha$.

Also, since $\beta \geqq 2$, $\beta\beta \geqq \beta \times 2$.

Therfor $\beta\beta > \alpha$.

533. Theorem. *If $\alpha > 2$ and we divide α successivly by each number of the natural series of increasing positiv integers, we must after a time come to a divisor β, wher $1 < \beta < \alpha$, which givs an exact or lower approximate quotient φ, such that $\varphi < \beta$.*

Let β be the integer one greater than the exact or lower approximate quotient obtaind by dividing α by 2.

Then $1 < \beta < \alpha$ and $\beta\beta > \alpha$.

Now divide α by β and let φ be the exact or lower approximate quotient obtaind.

Then $\varphi\beta \leqq \alpha$. (1)

Now suppose $\varphi \geqq \beta$.

Then $\varphi\beta \geqq \beta\beta$.

Hence $\varphi\beta > \alpha$,

which contradicts (1).

Hence $\varphi \neq \beta$ and $\varphi \ngtr \beta$.

Therfor $\varphi < \beta$.

We have shown, therfor, that ther is an integer β, which satisfies the requirements.

CHAPTER VII.

FACTORS.

534. We hav seen in § 487 that, if an integer has a given factor, it has also as a factor the opposit of the given factor. Moreover, no integer has zero as a factor (§ 427). We will therfor in what follows confine our attention chiefly to positiv factors.

535. Theorem. *The number of positiv factors of any integer, except zero, is not greater than its numerical value* (1).

For we hav seen that no factor is numerically greater than the integer itself.

If a is the given integer, its positiv factors must therfor be among the numbers $1, 2, 3, \cdots, |a| - 2, |a| - 1, |a|$, the number of which is $|a|$.

536. Hence the number of positiv factors of a given integer, not zero, is definit, or limited.

537. Every integer, not zero, has unity and its own numerical value as factors. But not every integer has other positiv factors. An example of the latter class is the number 7, as may be seen by dividing 7 (§ 505) by all the positiv integers less than it.

538. Definition. An integer that has no positiv fac-

(1) See Tannery, p. 111.

tor except its own numerical value, if this is positiv, and unity is calld a **prime** integer. All other integers ar **composit** integers and ar said to be **factorable.**

Zero is composit and unity is prime, for zero is divisible by every integer except itself, and unity has no positiv factor except itself.

We may determin the primeness or compositness of any given integer numerically greater than one by dividing it successivly by all the positiv integers numerically less than it, stopping if we come to an exact divisor greater than unity.

Thus we find that 1, 2, 3, 5, 7 ar prime integers and 4, 6, 8, 9 composit.

539. Theorem. *If the integer* α *is prime,* $-\alpha$ *is prime; if* α *is composit,* $-\alpha$ *is composit.*

For α and $-\alpha$ hav the same positiv factors.

540. Theorem. *If a prime integer* α *is divisible by an integer* β, *wher* $|\beta| \neq 1$, $|\alpha| = |\beta|$.

For, since α is prime, the only positiv factors it has ar $|\alpha|$ and 1.

Since α is divisible by β, $|\beta|$ must be a factor of α.

Therfor, since $|\beta| \neq 1$, $|\beta| = |\alpha|$.

541. Theorem. *Every composit integer, except zero, has a positiv prime factor numerically less than it and different from unity* (1).

For every composit integer α, except zero, has one or more positiv factors numerically less than it and different from unity.

---

(1) See Tannery, p. 127.

The smallest of these, say β, must be prime. For, if β had a positiv factor γ less than it and different from unity, α would also hav γ as a factor. § 460.

Then β would not be the smallest of the factors of α which are numerically less than it and different from unity.

Therfor β is prime.

542. In virtue of the last theorem, when using the process of § 538 to determin whether a given integer is prime or composit, we need not divide by any positiv integer that we know to be composit.

For, if the given integer is divisible by a composit positiv integer, it is also divisible by the positiv prime factors of this composit integer. § 460.

Hence, if not divisible by the prime factors, it is not divisible by the composit integer.

543. Theorem. *If* ν *is a composit integer different from zero, and* α *an integer such that* $|\nu| < \alpha\alpha$, *then* ν *must hav a positiv prime factor less than* $|\alpha|$ (1).

Since ν is composit and not zero, it has a positiv prime factor β, such that $1 < \beta < |\nu|$.

Hence $\nu = \beta\gamma$, wher $1 < |\gamma| < |\nu|$. § 501.

Hence, since $|\nu| < \alpha\alpha$, $|\beta\gamma| < \alpha\alpha$.

Therfor either β or γ is numerically less than α. § 408.

If $|\beta| < |\alpha|$, the theorem is proved.

In case $|\gamma| < |\alpha|$, γ is either prime or composit.

---

(1) See V. A. Le Besgue, "Théorie des Nombres," 1862, p. 48.

If γ is prime, then, since γ, and hence $|\gamma|$, is a factor of ν, the theorem is proved.

If γ is composit, it has a positiv prime factor δ, such that $1 < \delta < |\gamma|$.

Then ν is also divisible by δ (§ 460), which is less than $|\alpha|$ (§ 193), and the theorem is proved.

544. Theorem. *If ν is an integer different from zero, which has no positiv prime factor less than $|\alpha|$, and if $|\nu| < \alpha\alpha$, then ν is prime* (1).

This follows immediately from § 543.

545. Formation of a Table of Prime Positiv Integers (2). The method of § 538 is cumbersome, if used to find all the prime positiv integers from 1 to a given large number.

The following method is comparativly rapid.

Let the given number be ν and write the natural series of numbers $1, 2, 3, \cdots, \nu$.

To determin the prime numbers in this series we will cross out the composit numbers. This may be done by successivly crossing out the multiples, other than unit mutiples, of positiv integers which we know to be prime, other than unity. § 541.

Now we know that 2 is prime. § 538.

Let us cross out, then, the multiples of 2, 2×2, 2×3, 2×4, $\cdots$

The first number after 2 of those then remaining in the series is 3.

(1) See Tannery, p. 128.

(2) See Tannery, p. 128.

Now none of the numbers left in the series and less than 3×3 is divisible by any prime positiv integer less than 3, except 1.

All these numbers ar therfor prime. § 544.

Thus 3, 5, 7 ar prime.

Next we cross out the multiples of 3, 3×3, 3×4, $3 \times 5, \cdots$

The next number after 3 in the series now left is 5.

By the same reasoning as befor all numbers left in the series and less than 5×5 ar prime.

Next we cross out the multiples of 5, 5×5, 5×6, $5 \times 7, \cdots$

And so on.

If, after crossing out the multiples of the prime number γ, ther remain after γ in the series one or more numbers, the first of which is δ, δ is prime and we next cross out the multiples $\delta \times \delta$, $\delta(\delta + 1)$, $\delta(\delta + 2)$, $\cdots$, if these ar in the series. If $\delta \times \delta > \nu$, the table is complete. All the numbers then remaining in the series ar prime.

This method of forming a table of prime numbers is calld the **Siv of Eratosthenes.**

546. Theorem. *If, when a positiv integer* ν *is divided successivly by all the prime positiv integers from 2 to* δ, *wher* δ *need not be prime, none of the divisions is exact, and* δ *givs a lower quotient* φ *equal to or less than the prime number* ε *next greater than* δ, *the given integer is prime* (1).

(1) See Tannery, p. 130.

To prove this, we hav $\nu < (\varphi + 1)\delta$. § 512.

Now, if $\varphi < \varepsilon$, $\varphi + 1 \gtreqless \varepsilon$.

Hence, since $\delta < \varepsilon$, $(\varphi + 1)\delta < \varepsilon\varepsilon$. §§ 349, 350.

Therfor $\nu < \varepsilon\varepsilon$.

If $\varphi = \varepsilon$, let ρ be the remainder corresponding to the quotient φ.

Then, since $\varphi = \varepsilon > \delta > \rho$, $\varphi > \rho$.

Hence, if ν is divided by φ, the lower quotient will be δ. § 527.

Therfor $\nu < (\delta + 1)\varphi$.

Since $\delta < \varepsilon$, $\delta + 1 \gtreqless \varepsilon$.

Hence, since $\varphi = \varepsilon$, $(\delta + 1)\varphi \gtreqless \varepsilon\varepsilon$. §§ 331, 349.

Therfor $\nu < \varepsilon\varepsilon$.

In either case, then, $\nu < \varepsilon\varepsilon$.

Therfor, since ν has no positiv prime factor less than ε, ν is prime.

547. Theorem. *If, when a positiv integer ν is divided successivly by all the prime positiv integers from 2 to δ, wher δ need not be prime, none of the divisions is exact, and δ givs a lower quotient φ equal to or less than $\delta + 1$, the given integer is prime.*

This follows from § 546 as a particular case.

For, if $\varphi \gtreqless \delta + 1$, then, since $\delta + 1 \gtreqless \varepsilon$, $\varphi \gtreqless \varepsilon$.

548. By the help of the last two theorems we may use the table of prime positiv integers to advantage to determin whether a given positiv integer ν is prime or composit, even when ν is greater than any number in our table.

We divide ν successivly by the prime numbers of

our table. If we get an exact quotient greater than unity, ν is composit. If we get no exact quotient and come to a divisor which givs a lower quotient equal to or less than the next greater prime number in the table, the given number is prime. If we get no such quotient and try the last prime number η of our table as a divisor, then the given integer is prime if the quotient obtaind is equal to or less than $\eta + 1$. If this is not so, we try $\eta + 1$ as a divisor, then $\eta + 2$, $\eta + 3, \cdots$ until a divisor κ, less than ν, givs a quotient equal to or less than $\kappa + 1$.

We must come to such a divisor. For by § 533 we must find a divisor κ which givs an exact or lower approximate quotient less than κ.

If no quotient is exact, the given integer is prime. For, in dividing by all the prime numbers from 2 to η and all the integers $\eta + 1$, $\eta + 2$, $\eta + 3, \cdots, \kappa$, we divide by all the prime numbers from 2 to κ.

549. Theorem. *Having given any composit positiv integer, a set of positiv prime factors, all different from unity, can be found, of which it is the product* ([1]).

Consider the number $\mathfrak{C}$. Dividing by 2, we find that $\mathfrak{C} = 2 \times 6$. Dividing the quotient 6 by 2, we find that $6 = 2 \times 3$. Hence $\mathfrak{C} = 2 \times 2 \times 3$.

Similarly let ν be any given composit positiv integer. Then ν has some positiv prime factor less than it and different from unity. Let a be the first of the prime

---

([1]) See Tannery, p. 131; Chrystal, p. 38; Legendre, p. 5.

numbers 2, 3, 5, $\cdots$ by which ν is divisible, the quotient being φ_1. Then $\nu = a\varphi_1$.

If φ_1 is also composit and divisible by a, let φ_2 be the quotient, so that $\varphi_1 = a\varphi_2$ and $\nu = aa\varphi_2$.

Continuing this process we must come at last to a quotient φ_a which is not divisible by a.

For, let the series of quotients be $\varphi_1, \varphi_2, \varphi_3, \cdots$

Then $\nu > \varphi_1 > \varphi_2 > \varphi_3 > \cdots$ § 501.

As each quotient is less than the preceding and all ar less than ν, ther cannot be more than $\nu - 1$ of them. We must, then, at last come to a quotient φ_a which is not divisible by a.

Then we hav $\nu = (aaa \cdots a \text{ times})\, \varphi_a$.

In the same way we may find

$\varphi_a = (\beta\beta\beta \cdots b \text{ times})\chi_b,\ \chi_b = (\gamma\gamma\gamma \cdots c \text{ times})\psi_c$, etc.

Then

$\nu = (aaa \cdots a \text{ times})\, (\beta\beta\beta \cdots b \text{ times})\, (\gamma\gamma\gamma \cdots c \text{ times}) \cdots$

None of the quotients $\chi_1, \chi_2, \cdots, \chi_b$ is divisible by a. For then φ_a would be divisible by a. § 461.

Similarly none of the quotients $\psi_1, \psi_2, \cdots, \psi_c$ is divisible by a or β; and so on.

The whole series of quotients $\varphi_1, \varphi_2, \cdots, \varphi_a, \chi_1, \chi_2, \cdots, \chi_b, \psi_1, \psi_2, \cdots, \psi_c, \cdots$ must hav an end, the last quotient being a prime number different from unity, since, for the same reasons as given above, ther ar not more than $\nu - 1$ quotients in all.

We hav, then, found a set of positiv prime factors, of which ν is the product.

550. Definition. The process described above for

finding a set of positiv prime factors different from unity, of which a given composit positiv integer is the product, is calld **resolution into prime factors.**

In this sense we cannot properly speak of resolving a prime positiv integer into prime factors. We may, however, regard it as already resolvd.

551. Definition. A **common factor** of two or more integers in an integer that is a factor of each of them.

Thus 3 is a common factor of 6 and 9; 5 is a common factor of 5 and ϒ.

552. Theorem. *The number of positiv common factors of two or more integers, of which at least one is not zero, is not greater than the numerical value of the numerically smallest of those which ar not zero* (1).

Let the number of positiv common factors be n and let a be the numerically smallest of the integers which ar not zero.

Then, since the n positiv common factors ar factors of a, $n \ngtr |a|$. § 535.

553. Theorem. Thus *the number of positiv common factors of two or more integers, of which at least one is not zero, is limited.*

554. Theorem. *The number of positiv common factors of two or more zero integers is unlimited.*

555. Definition. An integer a is **prime** (2) **relativly to,** or simply **prime to,** an integer β, when a and β hav no positiv common factor except unity.

---

(1) See Tannery, p. 112.

(2) See Chrystal, p. 38.

Thus 4 and 9 ar prime to each other.

The statement "α is prime to β" may be written $\alpha \perp\!\!\!\perp \beta$; the statement "$\alpha$ is not prime to β" may be written $\alpha \not\perp\!\!\!\perp \beta$.

556. Theorem. *No integer except positiv and negativ unity is prime to itself.*

557. Theorem. *If* $\alpha \perp\!\!\!\perp \beta$, *then* $\beta \perp\!\!\!\perp \alpha$.

558. Theorem. *If* $\alpha \not\perp\!\!\!\perp \beta$, *then* $\beta \not\perp\!\!\!\perp \alpha$.

559. Theorem. *If* $\alpha = \beta$ *and* $\beta \perp\!\!\!\perp \gamma$, *then* $\alpha \perp\!\!\!\perp \gamma$.

560. Theorem. *If* $\alpha \perp\!\!\!\perp \beta$ *and* $\beta = \gamma$, *then* $\alpha \perp\!\!\!\perp \gamma$.

561. Theorem. *If* $\alpha \perp\!\!\!\perp \beta$, *then* $\bar{\alpha} \perp\!\!\!\perp \beta$, $\alpha \perp\!\!\!\perp \bar{\beta}$, *and* $\bar{\alpha} \perp\!\!\!\perp \bar{\beta}$.

For α has the same factors as $\bar{\alpha}$, and β as $\bar{\beta}$.

562. Theorem. $\alpha \perp\!\!\!\perp 1$.

563. Theorem. *The only positiv integer to which zero is prime is unity.*

564. Theorem. *Any two different prime positiv integers ar prime to each other.*

565. Theorem. *If* $\alpha \perp\!\!\!\perp \beta$, *every factor of* α *is prime to every factor of* β (1).

For, if these factors had a positiv common factor other than unity, α and β would hav this factor also.

566. Theorem. *If* $\alpha \gtrdot \beta$ *wher* $|\beta| \neq 1$, *then* $\alpha \not\perp\!\!\!\perp \beta$.

For, if $\alpha \gtrdot \beta$, wher $|\beta| \neq 1$, α and β hav a positiv common factor, $|\beta|$, different from unity.

Therfor $\alpha \not\perp\!\!\!\perp \beta$.

567. Theorem. *If* $\alpha \perp\!\!\!\perp \beta$, *wher* $|\beta| \neq 1$, *then* $\alpha \not\gtrdot \beta$.

This follows immediately from the preceding.

---

(1) See Tannery, p. 117.

568. Theorem. *If* $\alpha \not\perp\!\!\!\perp \beta$, *wher* β *is prime, then* $\alpha \gtrdot \beta$.

By the hypothesis $\beta \neq 0$.

For zero is a composit integer.

Moreover $|\beta| \neq 1$.

For every integer is prime to unity.

Now, since $\alpha \not\perp\!\!\!\perp \beta$, α and β must hav some positiv common factor besides unity.

But the only positiv factors of β ar 1 and $|\beta|$.

Hence $|\beta| \lessdot \alpha$.

Therfor $\alpha \gtrdot \beta$.

569. Theorem. *If* $\alpha \not\gtrdot \beta$, *wher* β *is prime, then* $\alpha \perp\!\!\!\perp \beta$.

570. Theorem. *If* $|\alpha| > |\beta|$, *wher* α *is prime and* $\beta \neq 0$, *then* $\alpha \perp\!\!\!\perp \beta$.

Since α is prime, the only positiv factors it has ar $|\alpha|$ and 1.

Moreover, no positiv common factor of α and β can be numerically greater than β.

Hence unity is the only positiv common factor of α and β.

Therfor $\alpha \perp\!\!\!\perp \beta$.

571. Theorem. *If* $\alpha = \varphi\beta + \gamma$, *any common factor of* β *and* γ *is a common factor of* α *and* β, *and conversely* (1).

The direct theorem is proved easily from § 480. The converse follows from § 481, since, if $\alpha = \varphi\beta + \gamma$, $\gamma = \alpha - \varphi\beta$.

(1) See Tannery, p. 112; Chrystal, p. 39.

CHAPTER VIII.

GREATEST COMMON FACTOR.

572. Definition. We hav seen that the number of positiv common factors of two or more integers, of which at least one is not zero, is limited (§ 553). Among these factors ther is one, and only one, which is greater than all the others (§ 199). This is calld their **greatest common factor** (1).

It may be found by dividing each of the other integers by the factors of the numerically smallest of those which ar not zero.

Thus the greatest common factor of 8 and ℭ is 4.

Two or more zero integers hav every positiv integer as a common factor and hence hav no greatest common factor.

We will denote the greatest common factor of two integers α and β, which ar not both zero, by the symbol $\alpha \otimes \beta$, which may be read "α kor β."

The symbol $\alpha \otimes \beta$ will, then, be univalent, if α and β ar not both zero. The symbol $0 \otimes 0$, however, will be meaningless.

The integers α and β may be calld the **elements** of the greatest common factor $\alpha \otimes \beta$.

---

(1) See Tannery, p. 112.

573. Operation of Finding Greatest Common Factor. In § 572 a method for finding the greatest common factor of two or more integers, which ar not both zero, was explaind. The greatest common factor $\alpha \otimes \beta$ is therfor found by an operation.

574. Theorem. $\alpha \otimes \beta = \beta \otimes \alpha$, *provided* α *and* β *ar not both zero.*

This follows immediately from §§ 553, 199, 170.

This theorem is the **Commutativ Law for Finding Greatest Common Factor.**

575. Theorem. *If* $\alpha \gtrdot \beta$, $\alpha \otimes \beta = |\beta|$ (1).

For the greatest positiv factor of β is $|\beta|$.

And, since $\alpha \gtrdot \beta$, $|\beta|$ is also a factor of α.

Therfor $|\beta|$ is the greatest common factor of α and β.

576. Theorem. $\alpha \otimes \alpha = |\alpha|$, *provided* $\alpha \neq 0$.

For then $\alpha \gtrdot \alpha$. § 442.

577. Theorem. $\alpha \otimes 1 = 1$.

578. Theorem. $0 \otimes \alpha = |\alpha|$, *provided* $\alpha \neq 0$.

579. Theorem. *If* $\alpha \perp\!\!\!\perp \beta$, $\alpha \otimes \beta = 1$, *and conversely.*

580. Theorem.

$$(-\alpha) \otimes \beta = \alpha \otimes (-\beta) = (-\alpha) \otimes (-\beta) = \alpha \otimes \beta,$$

provided α *and* β *ar not both zero.*

581. Theorem. *If* $|\gamma|$ *is unity*, $\alpha \otimes \gamma = \beta \otimes \gamma$.

For then both $\alpha \otimes \gamma$ and $\beta \otimes \gamma$ ar unity.

582. Theorem. *If* $|\gamma|$ *is unity*, $\gamma \otimes \alpha = \gamma \otimes \beta$.

583. Theorem. *If* $\alpha = \beta$, $\alpha \otimes \gamma = \beta \otimes \gamma$, *provided* β *and* γ *ar not both zero.*

---

(1) See J. A. Serret, "Traité d'Arithmetique," 1852, p. 79.

For, if α and β ar equal, they ar the same integer.

Hence they hav the same greatest common factor with γ.

584. Theorem. *If* $\alpha=\beta$, $\gamma \otimes \alpha = \gamma \otimes \beta$, *provided* β *and* γ *ar not both zero.*

585. Theorem. *If* $\alpha=\beta$ *and* $\gamma=\delta$, $\alpha \otimes \gamma = \beta \otimes \delta$, *provided* β *and* γ *ar not both zero.*

586. Theorem. *If* $\alpha = \varphi\beta + \gamma$, *then* $\alpha \otimes \beta = \beta \otimes \gamma$, *provided* α *and* β *ar not both zero* (1).

For $\alpha \otimes \beta$ is a common factor of β and γ
and $\beta \otimes \gamma$ is a common factor of α and β. § 571.

Hence, if $\alpha \otimes \beta > \beta \otimes \gamma$, $\beta \otimes \gamma$ would not be the greatest common factor of β and γ.

And, if $\alpha \otimes \beta < \beta \otimes \gamma$, $\alpha \otimes \beta$ would not be the greatest common factor of α and β.

Therfor $\alpha \otimes \beta = \beta \otimes \gamma$.

587. Theorem. *If* $\alpha = \varphi\beta + \gamma$ *and* $\alpha \perp\!\!\perp \beta$, *then* $\beta \perp\!\!\perp \gamma$, *and conversely.*

588. Theorem. $1 + \alpha\beta$ *is prime to* α *and* β.

For α and β are prime to 1.

589. Theorem. $1 + \alpha\beta\gamma \cdots$ *is prime to* $\alpha, \beta, \gamma, \cdots$

590. Theorem. *Having given any number of prime positiv integers, all different from unity, another can be found.*

For, let $\alpha, \beta, \gamma, \cdots, \kappa$ be any series of prime positiv integers, all different from unity,

and let $$\lambda = \alpha\beta\gamma \cdots \kappa + 1.$$

---

(1) See Le Besgue, p. 32; Chrystal, p. 39.

Then $\lambda > 1$ and λ is not divisible by any of the integers $\alpha, \beta, \gamma, \cdots, \kappa$. §§ 589, 567.

Now λ is either prime or composit.

If λ is prime, the theorem is proved. For, since λ is not divisible by any of the integers $\alpha, \beta, \gamma, \cdots, \kappa$, it cannot equal any of them. § 447.

If λ is composit, resolv it into positiv prime factors. § 549.

These factors must all be different from the integers $\alpha, \beta, \gamma, \cdots, \kappa$, since λ is not divisible by any of these integers.

Therfor another prime positiv integer, different from unity, can be found.

591. Theorem. *The number of prime positiv integers is infinit*(1).

592. Theorem. *Having given any prime positiv integer, a greater prime positiv integer can be found.*

Let κ be any prime positiv integer.

From the natural series of numbers $2, 3, 4, \cdots, \kappa$ pick out the prime integers. Let these be $\alpha, \beta, \gamma, \cdots, \kappa$.

Then, by § 590, having given the series of prime positiv integers $\alpha, \beta, \gamma, \cdots, \kappa$, all different from unity, another can be found.

But this other cannot be less than κ, since all the prime positiv integers different from unity and less than κ ar included in the series $\alpha, \beta, \gamma, \cdots, \kappa$.

(1) See Euclid's Elements, Bk. IX, Prop. 20; Chrystal, p. 47; Tannery, p. 127.

Therfor this other prime positiv integer is greater than κ.

This theorem is equivalent to the statement that *ther is no greatest prime positiv integer.*

593. Theorem. *Having given two integers α and β, wher $\beta \neq 0$; if we divide α by β, calling the remainder ρ_1; then, if $\rho_1 \neq 0$, divide β by ρ_1, calling the remainder ρ_2; then, if $\rho_2 \neq 0$, divide ρ_1 by ρ_2, calling the remainder ρ_3; and so on; that is, if we divide each remainder by the succeeding remainder, provided that succeeding remainder is not zero; we must finally come to a zero remainder.*

For, by § 519, $|\beta| > |\rho_1| > |\rho_2| > |\rho_3| > \cdots$

Hence the remainders continually decrease in numerical value and ar all numerically less than β.

The number of remainders cannot, then, be greater than the number of integers in the series $|\beta| - 1$, $|\beta| - 2$, $|\beta| - 3$, $\cdots$, 3, 2, 1, 0, which is $|\beta|$.

We must, therfor, at least as early as in the $|\beta|$'th division, find the remainder zero.

594. Theorem. *Having given two integers α and β, wher $\beta \neq 0$; if we divide α by β, calling the remainder ρ_1; then β by ρ_1, calling the remainder ρ_2; then ρ_1, by ρ_2, calling the remainder ρ_3; and so on, until we come to the zero remainder; the greatest common factor of α and β is the numerical value of the divisor that givs the zero remainder.*

Call the zero remainder ρ_n, set $\alpha = \rho_{-1}$, $\beta = \rho_0$, and call the successiv quotients $\varphi_1, \varphi_2, \varphi_3, \cdots, \varphi_n$.

From the series of divisions we get the series of equalities

$$\begin{aligned} \alpha &= \varphi_1\beta + \rho_1, \quad \text{or} & \rho_{-1} &= \varphi_1\rho_0 + \rho_1, \\ \beta &= \varphi_2\rho_1 + \rho_2, & \rho_0 &= \varphi_2\rho_1 + \rho_2, \\ \rho_1 &= \varphi_3\rho_2 + \rho_3, & \rho_1 &= \varphi_3\rho_2 + \rho_3, \\ &\cdots\cdots & &\cdots\cdots \\ \rho_{n-2} &= \varphi_n\rho_{n-1} + \rho_n, & \rho_{n-2} &= \varphi_n\rho_{n-1} + \rho_n. \end{aligned}$$

§ 515.

Hence $\alpha \otimes \beta = \beta \otimes \rho_1 = \rho_1 \otimes \rho_2 = \rho_2 \otimes \rho_3 = \cdots$

$= \rho_{n-2} \otimes \rho_{n-1} = \rho_{n-1} \otimes \rho_n$ § 586.

$= \rho_{n-1} \otimes 0$ § 584.

$= |\rho_{n-1}|.$ § 578.

This method for finding the greatest common factor of two integers is calld the **Algorithm of Euclid** [1].

595. Theorem. *If* $\alpha \perp\!\!\!\perp \beta$, $|\rho_{n-1}| = 1$, *and conversely.*

596. As the greatest common factor of two integers is an integer, we may combine it with other integers by any of the signs $+$, $-$, $\times$, $:$, $\otimes$, provided that the first element of every quotient is divisible by the second, and that of two integers connected by the sign $\otimes$ not both ar zero. We will use parentheses to indicate the order in which the various operations ar performd.

Every such expression is univalent. For every sum, difference, product, quotient, and greatest common factor is univalent.

---

(1) See Euclid, Bk. VII, Prop. 2; Chrystal, p. 39; Tannery, p. 113; Dirichlet-Dedekind, p. 6.

597. Theorem. *If for one or more elements of a complex expression which contains integers connected by auy or all of the signs* $+$, $-$, $\times$, $:$, $\otimes$, *and in which the first element of every quotient is divisible by the second and of two integers connected by the sign* $\otimes$ *not both ar zero, equal integers ar substituted, the complex expression is unchanged.*

598. Theorem.

If $\alpha \neq 0$ *and* $\beta \neq 0$, $(\alpha\beta) \otimes (\alpha\gamma) = |\alpha|(\beta \otimes \gamma)$ (1).

To obtain $\beta \otimes \gamma$ we proceed by the algorithm of Euclid, obtaining the remainders $\rho_1, \rho_2, \rho_3, \cdots, \rho_{n-2}, \rho_{n-1}, \rho_n$. § 594.

Then $\rho_n = 0$ and $\beta \otimes \gamma = |\rho_{n-1}|$.

To obtain $(\alpha\beta) \otimes (\alpha\gamma)$ we proceed in the same way.

But by § 524 a possible series of remainders will be $\alpha\rho_1, \alpha\rho_2, \alpha\rho_3, \cdots, \alpha\rho_{n-2}, \alpha\rho_{n-1}, \alpha\rho_n$.

And, since $\rho_n = 0$, $\alpha\rho_n = 0$.

Therfor $(\alpha\beta) \otimes (\alpha\gamma) = |\alpha\rho_{n-1}| = |\alpha||\rho_{n-1}|$

$$= |\alpha|(\beta \otimes \gamma).$$

This theorem is the **Left-handed Distributiv Law for Multiplication and the Operation of Finding Greatest Common Factor.**

599. Theorem.

If $\beta \neq 0$ *and* $\gamma \neq 0$, $(\alpha\gamma) \otimes (\beta\gamma) = (\alpha \otimes \beta)|\gamma|$.

This theorem is the **Right-handed Distributiv Law for Multiplication and the Operation of Finding Greatest Common Factor.**

---

(1) See Tannery, p. 116.

600. The preceding two theorems ar frequently of help in obtaining the greatest common factor of two given integers. Any common factor that is known to exist may be removed and the greatest common factor of the remaining numbers found. If this greatest common factor be multiplied by the numerical value of the factor that was removed, the product will be the required greatest common factor.

601. Theorem. *If* $\alpha \neq 0$, $\beta \neq 0$, *and* $\beta \perp\!\!\perp \gamma$, *then* $(\alpha\beta) \otimes (\alpha\gamma) = |\alpha|$; *and conversely, if* $(\alpha\beta) \otimes (\alpha\gamma) = |\alpha|$, *then* $\beta \perp\!\!\perp \gamma$.

For the direct theorem, since $\beta \perp\!\!\perp \gamma$, $\beta \otimes \gamma = 1$.

Hence $(\alpha\beta) \otimes (\alpha\gamma) = |\alpha| \times 1 = |\alpha|$. § 598.

For the converse theorem,

we hav $(\alpha\beta) \otimes (\alpha\gamma) = |\alpha|(\beta \otimes \gamma)$. § 598.

But by the hypothesis $(\alpha\beta) \otimes (\alpha\gamma) = |\alpha|$.

Hence $|\alpha|(\beta \otimes \gamma) = |\alpha|$ and $\beta \otimes \gamma = 1$.

Therfor $\beta \perp\!\!\perp \gamma$.

602. Theorem. *If* $\alpha \gtrdot \gamma$ *and* $\beta \gtrdot \gamma$, *wher* $\beta \neq 0$, *then* $\alpha \otimes \beta \gtrdot \gamma$ *and* $(\alpha \otimes \beta) : |\gamma| = (\alpha : \gamma) \otimes (\beta : \gamma)$ (1).

We hav $\alpha = \varphi\gamma$ and $\beta = \chi\gamma$, wher $\chi \neq 0$.

Hence $\alpha \otimes \beta = (\varphi\gamma) \otimes (\chi\gamma) = (\varphi \otimes \chi)|\gamma|$.

§§ 585, 599.

Therfor $\alpha \otimes \beta \gtrdot \gamma$

and $(\alpha \otimes \beta) : |\gamma| = \varphi \otimes \chi = (\alpha : \gamma) \otimes (\beta : \gamma)$.

This theorem is the **Distributiv Law for Division and the Operation of Finding Greatest Common Factor**

(1) See Le Besgue, p. 32.

603. Theorem. *If* $\beta \neq 0$; *then, if* $|\theta| = \alpha \otimes \beta$, $\alpha \succ \theta$, $\beta \succ \theta$, *and* $\alpha : \theta \perp\!\!\perp \beta : \theta$; *and conversely, if* $\alpha \succ \theta$, $\beta \succ \theta$, *and* $\alpha : \theta \perp\!\!\perp \beta : \theta$, *then* $|\theta| = \alpha \otimes \beta$.

For the direct theorem,

$$(\alpha : \theta) \otimes (\beta : \theta) = (\alpha \otimes \beta) : |\theta| = 1. \quad \S\, 602.$$

Hence $\alpha : \theta \perp\!\!\perp \beta : \theta$.

For the converse theorem reverse the steps.

604. This theorem is equivalent to the following:

Theorem. *If* α *and* β *ar two integers, which ar not both zero; then, if* κ *is their greatest common factor, ther exist two integers* γ *and* δ, *prime to each other, such that* $\alpha = \gamma\kappa$ *and* $\beta = \delta\kappa$; *and conversely, if* $\alpha = \gamma\kappa$ *and* $\beta = \delta\kappa$, *wher* γ *and* δ *ar prime to each other and* κ *is positiv and* δ *not zero, then* κ *is the greatest common factor of* α *and* β.

605. Theorem. *The common factors of two integers, which ar not both zero, ar the same as the factors of their greatest common factor.*

For, by §§ 572, 460, every factor of the greatest common factor is a factor of each of the given integers.

And, by § 602, every common factor of the two given integers is a factor of their greatest common factor.

Hence the two groups of factors, first, the common factors of the two integers and, second, the factors of their greatest common factor, ar the same.

606. Theorem. *The integers α and β, wher $\beta \neq 0$, and all the remainders obtaind in finding their greatest common factor by the algorithm of Euclid can be exprest in the form $\lambda\alpha + \mu\beta$* [1].

We hav
$$\rho_{-1} = \varphi_1\rho_0 + \rho_1,$$
$$\rho_0 = \varphi_2\rho_1 + \rho_2,$$
$$\rho_1 = \varphi_3\rho_2 + \rho_3,$$
$$. \quad . \quad . \quad . \quad . \quad . \quad . \quad . \quad ,$$
$$\rho_{n-2} = \varphi_n\rho_{n-1} + \rho_n,$$
wher ρ_n is the last remainder, the zero remainder.

§ 594.

Now $\alpha = 1.\alpha - 0.\beta$,

or $\rho_{-1} = \lambda_{-1}\alpha + \mu_{-1}\beta$, wher $\lambda_{-1} = 1$ and $\mu_{-1} = -0$.

Again $\beta = -0.\alpha + 1.\beta$, or $\rho_0 = \lambda_0\alpha + \mu_0\beta$,

wher $\lambda_0 = -0$ and $\mu_0 = 1$.

Thus the first two of the numbers $\alpha, \beta, \rho_1, \rho_2, \rho_3, \cdots, \rho_n$ hav been exprest in the desired form.

Suppose we hav proved that two successiv ρ's, ρ_{m-2} and ρ_{m-1}, wher $m - 2 \gtreqless -1$ and $m - 1 \lesseqgtr n - 1$, that is, wher $0 < m < n + 1$, can be written in the form $\lambda\alpha + \mu\beta$, so that $\rho_{m-2} = \lambda_{m-2}\alpha + \mu_{m-2}\beta$

and
$$\rho_{m-1} = \lambda_{m-1}\alpha + \mu_{m-1}\beta.$$

Then, since
$$\rho_m = -\varphi_m\rho_{m-1} + \rho_{m-2},$$
by substituting in this equality the values of ρ_{m-1} and ρ_{m-2} taken from the preceding two equalities, we get

---

(1) See Chrystal, p. 44.

$$\rho_m = -\varphi_m(\lambda_{m-1}\alpha + \mu_{m-1}\beta) + (\lambda_{m-2}\alpha + \mu_{m-2}\beta)$$
$$= (-\varphi_m\lambda_{m-1} + \lambda_{m-2})\alpha + (-\varphi_m\mu_{m-1} + \mu_{m-2})\beta$$
$$= \lambda_m\alpha + \mu_m\beta,$$

wher $$\lambda_m = -\varphi_m\lambda_{m-1} + \lambda_{m-2}$$

and $$\mu_m = -\varphi_m\mu_{m-1} + \mu_{m-2}.$$

Thus we hav proved that, if two successiv remainders ρ_{m-2} and ρ_{m-1} can be written in the form $\lambda\alpha + \mu\beta$, the next remainder ρ_m can also be written in this form.

But we know that ρ_{-1} and ρ_0 can be written in the form $\lambda\alpha + \mu\beta$. Hence ρ_1 can be written in this form. Again, since ρ_0 and ρ_1 can be written in this form, ρ_2 can be. And so on. That is, all the ρ's can be written in this form. Therfor $\rho_\iota = \lambda_\iota\alpha + \mu_\iota\beta$, wher ι may hav any value from -1 to n.

The formulas $\lambda_m = -\varphi_m\lambda_{m-1} + \lambda_{m-2}$

and $$\mu_m = -\varphi_m\mu_{m-1} + \mu_{m-2}$$

show how, having given λ_{-1}, λ_0 and μ_{-1}, μ_0 and the series of φ's, the remaining λ's and μ's can be calculated.

Thus $$\lambda_1 = -\varphi_1\lambda_0 + \lambda_{-1}, \qquad \mu_1 = -\varphi_1\mu_0 + \mu_{-1},$$
$$\lambda_2 = -\varphi_2\lambda_1 + \lambda_0, \qquad \mu_2 = -\varphi_2\mu_1 + \mu_0,$$
. . . .

It will be noticed that the formulas by which λ_m and μ_m ar calculated from $\lambda_{m-1}, \lambda_{m-2}$ and μ_{m-1}, μ_{m-2} respectivly ar similar to that by which ρ_m is obtaind from ρ_{m-1} and ρ_{m-2}.

607. Definition. The method of proof just used to

show that ρ_ι can be written in the form $\lambda_\iota\alpha + \mu_\iota\beta$ for all possible values of ι is calld the **Method of Mathematical Induction.**

According to this method we prove that if a certain rule, in whose statement ther appears an integer η holds for a particular value of η, or for some specified number of successiv particular values, then it holds for the next value.

Then we show that it does hold for a particular value of η, or for the specified number of successiv particular values.

Hence it holds for the next value and therfor for all values that η can take, starting with and succeeding some particular value.

This result may be stated in the form of a theorem as follows:

Theorem. Law of Mathematical Induction. *If a certain statement, in which an integer η is involvd, holds for a particular value of η or for some specified number of successiv particular values, and, if it is proved that, if the statement is true for a particular value of η or for the specified number of successiv particular values, it holds for the next value, then the statement is true for all values that η can take, starting with and succeeding some particular value.*

This is so, because every integer has its place in the natural series of integers.

608. By the method of § 606 we can calculate as

many values as we please of the λ's and μ's in terms of the φ's, up to λ_n and μ_n.

Thus we hav

$$\lambda_{-1} = +1, \qquad \mu_{-1} = -0,$$
$$\lambda_0 = -0, \qquad \mu_0 = +1,$$
$$\lambda_1 = +1, \qquad \mu_1 = -\varphi_1,$$
$$\lambda_2 = -\varphi_2, \qquad \mu_2 = +(\varphi_2\varphi_1 + 1),$$
$$\lambda_3 = +(\varphi_3\varphi_2 + 1), \qquad \mu_3 = -(\varphi_3\varphi_2\varphi_1 + \varphi_3 + \varphi_1),$$
$$\cdots\cdots \qquad \cdots\cdots$$

It will be noticed that the signs with which the values of the λ's and μ's begin ar alternately plus and minus and that each λ has the opposit sign to the corresponding μ. This rule holds for all the λ's and μ's

For, since $\lambda_m = -\varphi_m\lambda_{m-1} + \lambda_{m-2}$, if λ_{m-2} has the plus sign and λ_{m-1} the minus sign, λ_m will hav the plus sign (§ 326). And if λ_{m-2} has the minus sign and λ_{m-1} the plus sign, λ_m will hav the minus sign.

Hence, since λ_{-1} has the plus sign and λ_0 the minus sign, λ_1 has the plus sign; since λ_0 has the minus sign and λ_1 the plus sign, λ_2 has the minus sign; and so on.

Similarly for the μ's.

Also, since λ_{-1} has the opposit sign to μ_{-1}, and the signs of the λ's and μ's alternate as we go down the series, the signs of the corresponding λ's and μ's will continue to be opposit.

609. Theorem. *If $\lambda_{\iota-1}$, λ_ι and $\mu_{\iota-1}$, μ_ι ar any pair of successiv λ's and the corresponding μ's in the series $\lambda_1, \lambda_2, \lambda_3, \cdots, \lambda_n$ and $\mu_1, \mu_2, \mu_3, \cdots, \mu_n$, given in § 606, then $\lambda_{\iota-1}\mu_\iota - \lambda_\iota\mu_{\iota-1} = \pm 1$, according as ι is even or odd.*

Set $\tau_\iota = \lambda_{\iota-1}\mu_\iota - \lambda_\iota\mu_{\iota-1}$, wher $-1 < \iota < n+1$.

Then $$\tau_0 = \lambda_{-1}\mu_0 - \lambda_0\mu_{-1}$$
$$= 1 \times 1 - 0 \times 0 = 1.$$

Now we hav $\lambda_\iota = -\varphi_\iota\lambda_{\iota-1} + \lambda_{\iota-2}$

and $$\mu_\iota = -\varphi_\iota\mu_{\iota-1} + \mu_{\iota-2},$$
which equalities hold when $0 < \iota < n+1$.

By multiplying the second of these equalities by $\lambda_{\iota-1}$ and the first by $\mu_{\iota-1}$, and subtracting, we get

$$\lambda_{\iota-1}\mu_\iota - \lambda_\iota\mu_{\iota-1} = -(\lambda_{\iota-2}\mu_{\iota-1} - \lambda_{\iota-1}\mu_{\iota-2}),$$

or $\tau_\iota = -\tau_{\iota-1}$, which holds if $0 < \iota < n+1$.

In this equality replacing ι by $\iota - 1$, we get $\tau_{\iota-1} = -\tau_{\iota-2}$, which holds if $0 < \iota - 1 < n+1$, or $1 < \iota < n+2$.

Hence $\tau_\iota = \tau_{\iota-2}$, or $\tau_{\iota-2} = \tau_\iota$, which holds when both $0 < \iota < n+1$ and $1 < \iota < n+2$, that is, when $1 < \iota < n+1$.

In the latter equality giving ι the values $2, 4, 6, \cdots$ in succession, we see that $\tau_0 = \tau_2 = \tau_4 = \tau_6 = \cdots$

Giving ι the values $3, 5, 7, \cdots$ in succession, we get $\tau_1 = \tau_3 = \tau_5 = \tau_7 = \cdots$

Now $\tau_0 = 1$. Hence $\tau_\iota = 1$, whenever ι is an even number greater than -1 and less than $n+1$.

Also, since $\tau_\iota = -\tau_{\iota-1}$, when $0 < \iota < n+1$,

$$\tau_1 = -\tau_0 = -1.$$

Hence $\tau_\iota = -1$, whenever ι is an odd number greater than -1 and less than $n+1$.

610. Theorem. *If α and β ar two integers, which ar* *not both zero, two integers* λ *and* μ *can be found, such that* $\lambda\alpha + \mu\beta = \alpha \otimes \beta$.

We hav $\alpha \otimes \beta = |\rho_{n-1}|$.

And $\rho_{n-1} = \lambda_{n-1}\alpha + \mu_{n-1}\beta$.

Hence

$$|\rho_{n-1}| = \pm(\lambda_{n-1}\alpha + \mu_{n-1}\beta) = \pm\lambda_{n-1}\alpha \pm \mu_{n-1}\beta,$$

the upper or the lower sign being used according as ρ_{n-1} is positiv or negativ.

If, then, ρ_{n-1} is positiv,

we hav $\lambda_{n-1}\alpha + \mu_{n-1}\beta = \alpha \otimes \beta$.

If ρ_{n-1} is negativ,

$$-\lambda_{n-1}\alpha - \mu_{n-1}\beta = \alpha \otimes \beta.$$

In the first case setting λ_{n-1}, μ_{n-1} respectivly equal to λ, μ and in the second case setting λ_{n-1}, μ_{n-1} equal to $-\lambda$, $-\mu$, we hav for each case $\lambda\alpha + \mu\beta = \alpha \otimes \beta$.

611. Theorem. *If* α *and* β *ar two integers prime to each other, two integers* λ *and* μ *can be found, such that* $\lambda\alpha + \mu\beta = 1$.

According to the hypothesis α and β cannot both be zero, because zero is not prime to itself. § 556.

Hence this theorem follows from the previous one and § 579.

612. Theorem. *If* $\alpha + \beta = 1$, *then* $\alpha \perp\!\!\perp \beta$.

For, since $\alpha + \beta = 1$, any common factor of α and β is a factor of 1. § 478.

613. Theorem. *If* $\alpha + \beta = 1$ *and* γ *and* δ *ar factors of* α *and* β *respectively,* $\gamma \perp\!\!\perp \delta$.

This follows easily from the preceding theorem and § 460.

614. Theorem. *If $\lambda\alpha + \mu\beta = 1$, each of the integers λ, α is prime to each of the integers μ, β.*

This follows from the preceding theorem, since $\lambda\alpha \not> \alpha$, $\mu\beta \not> \beta$, etc.

615. Theorem. *In the series $\lambda_{-1}, \lambda_0, \lambda_1, \cdots, \lambda_n$ and $\mu_{-1}, \mu_0, \mu_1, \cdots, \mu_n$, given in § 606, each λ is prime to the succeeding λ, each μ to the succeeding μ, and each λ to the corresponding μ.*

This follows easily from the preceding theorem and § 609.

616. Theorem. *If $\lambda\alpha + \mu\beta = \alpha \otimes \beta$, then $\lambda \perp\!\!\!\perp \mu$.*

We hav $\alpha = \gamma\kappa$ and $\beta = \delta\kappa$, wher $\kappa = \alpha \otimes \beta$. § 604.

Hence $\qquad \lambda\gamma\kappa + \mu\delta\kappa = \kappa$.

Hence $\qquad \lambda\gamma + \mu\delta = 1$.

Therfor $\qquad \lambda \perp\!\!\!\perp \mu$. § 614.

This theorem might be stated. *In whatever way $\alpha \otimes \beta$ is exprest in the form $\lambda\alpha + \mu\beta$, λ and μ ar prime to each other.*

617. The most important particular case of the preceding investigation, §§ 593–611, is the case when α and β ar positiv. Let us consider this case.

Suppose $\alpha > \beta$ and that the quotients $\varphi_1, \varphi_2, \cdots, \varphi_n$ ar all lower approximate quotients, except φ_n, which is exact, so that the remainders $\rho_1, \rho_2, \cdots, \rho_n$ ar all positiv, except ρ_n, which is zero.

Then the greatest common factor of α and β is ρ_{n-1}, which is unity, if α and β ar prime to each other.

The quotients $\varphi_1, \varphi_2, \cdots, \varphi_n$ ar all positiv. § 526.

Hence the values of $\lambda_1, \lambda_2, \cdots, \lambda_n$ (§ 608) ar alternately positiv and negativ, the positiv values corresponding to odd subscripts.

The values of $\mu_1, \mu_2, \cdots, \mu_n$ ar also alternately positiv and negativ, the positiv values corresponding to even subscripts.

Of any λ and the corresponding μ, one is positiv and the other negativ.

Let us now put $\lambda_1, \lambda_3, \lambda_5, \cdots$ respectivly equal to $\nu_1, \nu_3, \nu_5, \cdots$; $\lambda_2, \lambda_4, \lambda_6, \cdots$ equal to $-\nu_2, -\nu_4, -\nu_6, \cdots$; $\mu_1, \mu_3, \mu_5, \cdots$ equal to $-\xi_1, -\xi_3, -\xi_5, \cdots$; $\mu_2, \mu_4, \mu_6, \cdots$ equal to $\xi_2, \xi_4, \xi_6, \cdots$

Then all the ν's and ξ's will be positiv and we will hav

$$\nu_1 = 1, \qquad \xi_1 = \varphi_1,$$
$$\nu_2 = \varphi_2, \qquad \xi_2 = \varphi_2\varphi_1 + 1,$$
$$\nu_3 = \varphi_3\varphi_2 + 1, \qquad \xi_3 = \varphi_3\varphi_2\varphi_1 + \varphi_3 + \varphi_1,$$
$$\cdots\cdots, \qquad \cdots\cdots,$$

$$\rho_1 = \nu_1\alpha - \xi_1\beta,$$
$$\rho_2 = -\nu_2\alpha + \xi_2\beta,$$
$$\rho_3 = \nu_3\alpha - \xi_3\beta,$$
$$\cdots\cdots,$$
$$\rho_m = \pm(\nu_m\alpha - \xi_m\beta), \text{ according as}$$

m is odd or even.

Also
$$\nu_m = \varphi_m\nu_{m-1} + \nu_{m-2},$$
$$\xi_m = \varphi_m\xi_{m-1} + \xi_{m-2}, \text{ if } m > 2.$$

For, if m is odd, the formula $\lambda_m = -\varphi_m\lambda_{m-1} + \lambda_{m-2}$ givs
$$\nu_m = (-\varphi_m)(-\nu_{m-1}) + \nu_{m-2},$$
or
$$\nu_m = \varphi_m\nu_{m-1} + \nu_{m-2}.$$

If m is even, the same formula givs

$$-\nu_m = -\varphi_m\nu_{m-1} - \nu_{m-2},$$

or
$$\nu_m = \varphi_m\nu_{m-1} + \nu_{m-2},$$

the same as in the other case.

Similarly the formula for ξ_m is proved.

We can now obtain from the theorems of §§ 606, 609–611 the following theorems for this particular case.

618. Theorem. *If α and β ar positiv integers and $\alpha > \beta$, all the remainders obtaind in finding their greatest common factor by the algorithm of Euclid can be exprest in the form $\pm(\nu\alpha - \xi\beta)$, wher ν and ξ ar positiv integers, the upper sign being used for the odd remainders, the lower for the even remainders.*

619. Theorem. *If ν_{m-1}, ν_m and ξ_{m-1}, ξ_m ar any pair of successiv ν's and the corresponding ξ's in the series $\nu_1, \nu_2, \cdots, \nu_n$ and $\xi_1, \xi_2, \cdots, \xi_n$, $\nu_{m-1}\xi_m - \nu_m\xi_{m-1} = \pm 1$, according as m is even or odd.*

For, if m is even, the equality $\lambda_{m-1}\mu_m - \lambda_m\mu_{m-1} = +1$ givs
$$\nu_{m-1}\xi_m - (-\nu_m)(-\xi_{m-1}) = +1,$$
or
$$\nu_{m-1}\xi_m - \nu_m\xi_{m-1} = +1.$$

Similarly the other case is proved.

620. Theorem. *In the series $\nu_1, \nu_2, \cdots, \nu_n$ and $\xi_1, \xi_2, \cdots, \xi_n$ each ν is prime to the succeeding ν, each ξ to the succeeding ξ, and each ν to the corresponding ξ.*

621. Theorem. *If α and β ar positiv integers, two positiv integers ν and ξ can be found, such that $\nu\alpha - \xi\beta = \pm\alpha \otimes \beta$* [1].

---

[1] See Le Besgue, p. 37; Chrystal, p. 45.

622. Theorem. *If α and β ar positiv integers prime to each other, two positiv integers ν and ξ can be found, such that $\nu\alpha - \xi\beta = \pm 1$.*

623. We will now leave this particular case and continue the general investigation.

Theorem. *If $\theta\alpha \gtrdot \beta$, wher $\alpha \perp\!\!\perp \beta$, then $\theta \gtrdot \beta$* [1].

Since $\alpha \perp\!\!\perp \beta$, we can find λ and μ, such that

$$1 = \lambda\alpha + \mu\beta. \qquad \S\ 611.$$

Multiplying by θ, $\theta = \lambda\theta\alpha + \mu\theta\beta$.

Since $\theta\alpha \gtrdot \beta$, $\theta\alpha = \varphi\beta$.

Substituting $\varphi\beta$ for $\theta\alpha$ in the preceding equality,

$$\theta = \lambda\varphi\beta + \mu\theta\beta = (\lambda\varphi + \mu\theta)\beta.$$

Therfor $\theta \gtrdot \beta$.

624. Theorem. *If $\alpha \perp\!\!\perp \beta$, any common factor of $\theta\alpha$ and β must be a factor of θ* [2].

625. Theorem. *If $\theta \perp\!\!\perp \beta$, $(\alpha\theta) \otimes \beta = \alpha \otimes \beta$, provided α and β ar not both zero* [3].

For $\alpha \otimes \beta$ is a common factor of $\alpha\theta$ and β. § 460.

And $(\alpha\theta) \otimes \beta$ is a common factor of α and β. § 624.

Hence, if $(\alpha\theta) \otimes \beta > \alpha \otimes \beta$, $\alpha \otimes \beta$ would not be the greatest common factor of α and β.

And, if $(\alpha\theta) \otimes \beta < \alpha \otimes \beta$, $(\alpha\theta) \otimes \beta$ would not be the greatest common factor of $\alpha\theta$ and β.

Therfor $(\alpha\theta) \otimes \beta = \alpha \otimes \beta$.

---

(1) See Chrystal, p. 41; Tannery, p. 119; Lucas, p. 339. This theorem was known to Euclid.

(2) See Dirichlet-Dedekind, p. 8.

(3) See Tannery, p. 118.

626. Theorem. *If* $\alpha \mathrel{\dot{>}} \theta$ *and* $\theta \perp\!\!\!\perp \beta$, *then* $(\alpha : \theta) \otimes \beta = \alpha \otimes \beta$, *provided* α *and* β *ar not both zero.*

Since $\alpha \mathrel{\dot{>}} \theta$, $\alpha = \varphi\theta$.

Hence $\alpha \otimes \beta = (\varphi\theta) \otimes \beta$ § 583.

$= \varphi \otimes \beta$ § 625.

$= (\alpha : \theta) \otimes \beta.$ § 583.

627. Theorem. $(\alpha\beta) \otimes \gamma$ *is divisible by* $\beta \otimes \gamma$.

For we hav $\beta = \delta\kappa$ and $\gamma = \varepsilon\kappa$, wher $\kappa = \beta \otimes \gamma$. § 604.

Hence $(\alpha\beta) \otimes \gamma = (\alpha\delta\kappa) \otimes (\varepsilon\kappa) = [(\alpha\delta) \otimes \varepsilon]\kappa$. § 599.

Therfor $(\alpha\beta) \otimes \gamma \mathrel{\dot{>}} \kappa$. § 409.

Or $(\alpha\beta) \otimes \gamma \mathrel{\dot{>}} \beta \otimes \gamma$.

628. Theorem. *If* $\nu \mathrel{\dot{>}} \theta\alpha$ *and* $\nu \mathrel{\dot{>}} \theta\beta$, *wher* $\nu \neq 0$ *and* $\alpha \perp\!\!\!\perp \beta$, *then* $\nu \mathrel{\dot{>}} \theta\alpha\beta$ *and* $|\nu| : |\theta\alpha\beta| = (\nu : (\theta\alpha)) \otimes (\nu : (\theta\beta))$; *and conversely, if* $\nu \mathrel{\dot{>}} \theta\alpha\beta$, *then* $\nu \mathrel{\dot{>}} \theta\alpha$, $\nu \mathrel{\dot{>}} \theta\beta$, *and, if* $|\nu| : |\theta\alpha\beta| = (\nu : (\theta\alpha)) \otimes (\nu : (\theta\beta))$, *then* $\alpha \perp\!\!\!\perp \beta$.

For the direct theorem,

we hav $\nu = \varphi\theta\alpha$ and $\nu = \chi\theta\beta$.

Hence $\varphi\theta\alpha = \chi\theta\beta$.

Hence $\varphi\alpha = \chi\beta$. § 339.

That is, $\varphi\alpha \mathrel{\dot{>}} \beta$.

Hence, since $\alpha \perp\!\!\!\perp \beta$, $\varphi \mathrel{\dot{>}} \beta$. § 623.

That is, $\varphi = \lambda\beta$.

Hence $\nu = \lambda\theta\alpha\beta$.

Therfor $\nu \mathrel{\dot{>}} \theta\alpha\beta$ and $\nu : (\theta\alpha\beta) = \lambda$, $\nu : (\theta\alpha) = \lambda\beta$, $\nu : (\theta\beta) = \lambda\alpha$.

Hence $|\nu| : |\theta\alpha\beta| = |\lambda|$
and
$(\nu : (\theta\alpha)) \otimes (\nu : (\theta\beta)) = (\lambda\beta) \otimes (\lambda\alpha)$ § 585.
$= |\lambda|$ § 601.

Therfor $|\nu| : |\theta\alpha\beta| = (\nu : (\theta\alpha)) \otimes (\nu : (\theta\beta))$.

For the converse theorem reverse the steps of the latter part of the proof.

629. Theorem. *If $\nu \mathrel{\dot{>}} \alpha$ and $\nu \mathrel{\dot{>}} \beta$, wher $\nu \neq 0$ and $\alpha \perp\!\!\perp \beta$, then $\nu \mathrel{\dot{>}} \alpha\beta$ and $|\nu| : |\alpha\beta| = (\nu : \alpha) \otimes (\nu : \beta)$; and conversely, if $\nu \mathrel{\dot{>}} \alpha\beta$, then $\nu \mathrel{\dot{>}} \alpha, \nu \mathrel{\dot{>}} \beta$, and, if $|\nu| : |\alpha\beta| = (\nu : \alpha) \otimes (\nu : \beta)$, then $\alpha \perp\!\!\perp \beta$.*

This theorem is a particular case of the preceding theorem, the case when $\theta = 1$.

The theorem of § 579 is a particular case of this theorem, the case when $\nu = \alpha\beta$.

630. Theorem. *If $\nu \mathrel{\dot{>}} \alpha$ and $\nu \mathrel{\dot{>}} \beta$, wher $\nu \neq 0$, then $\nu \mathrel{\dot{>}} (\alpha\beta) : (\alpha \otimes \beta)$ and $|\nu| : (|\alpha\beta| : (\alpha \otimes \beta)) = (\nu : \alpha) \otimes (\nu : \beta)$.*

We hav $\alpha = \kappa\gamma$ and $\beta = \kappa\delta$,
wher $\alpha \otimes \beta = \kappa$ and $\gamma \perp\!\!\perp \delta$. § 604.

Since $\nu \mathrel{\dot{>}} \alpha$ and $\nu \mathrel{\dot{>}} \beta$, $\nu \mathrel{\dot{>}} \kappa\gamma$ and $\nu \mathrel{\dot{>}} \kappa\delta$. § 459.

Hence $\nu \mathrel{\dot{>}} \kappa\gamma\delta$
and $|\nu| : |\kappa\gamma\delta| = (\nu : (\kappa\gamma)) \otimes (\nu : (\kappa\delta))$. § 628.

Moreover $\alpha\beta = (\kappa\gamma)(\kappa\delta) = (\kappa\gamma\delta)\kappa$. §§ 331, 381.

Therfor $\alpha\beta \mathrel{\dot{>}} \kappa$ and $(\alpha\beta) : \kappa = \kappa\gamma\delta$. § 414.

Hence $|\kappa\gamma\delta| = |(\alpha\beta) : \kappa| = |\alpha\beta| : \kappa$ §§ 396, 496.
$= |\alpha\beta| : (\alpha \otimes \beta)$. § 459.

Therfor $\nu \mathrel{\dot{>}} (\alpha\beta) : \kappa$ and
$|\nu| : (|\alpha\beta| : (\alpha \otimes \beta)) = (\nu : \alpha) \otimes (\nu : \beta)$. §§ 459, 585.

631. Theorem. *If* $\alpha \perp\!\!\perp \beta$ *and* $\alpha \perp\!\!\perp \gamma$, *then* $\alpha \perp\!\!\perp \beta\gamma$; *and conversely, if* $\alpha \perp\!\!\perp \beta\gamma$, *then* $\alpha \perp\!\!\perp \beta$ *and* $\alpha \perp\!\!\perp \gamma$ (1).

For the direct theorem, since $\alpha \perp\!\!\perp \beta$, α and β cannot both be zero.

Hence, since $\alpha \perp\!\!\perp \gamma$, $\alpha \otimes (\beta\gamma) = \alpha \otimes \beta$ § 622.

$= 1.$ § 579.

Therfor $\alpha \perp\!\!\perp \beta\gamma$. § 579.

For the converse theorem,

since $\alpha \perp\!\!\perp \beta\gamma$, $\alpha \otimes (\beta\gamma) = 1.$

Also α and β cannot both be zero.

For then α and $\beta\gamma$ would be zero, which cannot be, since $\alpha \perp\!\!\perp \beta\gamma$.

But $\alpha \otimes (\beta\gamma)$ is a multiple of both $\alpha \otimes \beta$ and $\alpha \otimes \gamma$. § 627.

Hence, if either $\alpha \otimes \beta$ or $\alpha \otimes \gamma$ were not unity, $\alpha \otimes (\beta\gamma)$ would not be unity.

Therfor both $\alpha \otimes \beta$ and $\alpha \otimes \gamma$ ar unity.

Therfor $\alpha \perp\!\!\perp \beta$ and $\alpha \perp\!\!\perp \gamma$.

632. Theorem. *If* α *is prime to each of two or more integers* $\beta, \gamma, \delta, \cdots$, *it is prime to their product* $\beta\gamma\delta \cdots$; *and conversely, if* α *is prime to the product* $\beta\gamma\delta \cdots$, *then it is prime to each of those integers* (2).

We hav already proved the theorem for the case of two integers.

Suppose ther ar more than two.

Then, for the direct theorem, we hav first $\alpha \perp\!\!\perp \beta\gamma$. § 631.

(1) See Tannery, p. 119.

(2) See Chrystal, p. 41.

Hence, since $\alpha \perp\!\!\!\perp \beta\gamma$ and $\alpha \perp\!\!\!\perp \delta$, $\alpha \perp\!\!\!\perp (\beta\gamma)\delta$. § 631.

And so on.

For the converse theorem, since $\alpha \perp\!\!\!\perp \beta(\gamma\delta \cdots)$, $\alpha \perp\!\!\!\perp \beta$ and $\alpha \perp\!\!\!\perp \gamma\delta \cdots$ § 631.

Similarly α can be proved prime to each of the other integers.

633. Theorem. *If each of the integers α, β, γ, $\cdots$ is prime to each of the integers λ, μ, ν, $\cdots$, the product $\alpha\beta\gamma \cdots$ is prime to the product $\lambda\mu\nu \cdots$; and conversely, if the product $\alpha\beta\gamma \cdots$ is prime to the product $\lambda\mu\nu \cdots$, then each of the integers α, β, γ, $\cdots$ is prime to each of the integers λ, μ, $\nu \cdots$* (1).

For the direct theorem, we hav first $\alpha \perp\!\!\!\perp \lambda\mu\nu \cdots$ § 632.

Similarly $\beta \perp\!\!\!\perp \lambda\mu\nu \cdots$

And so on.

Therfor $\alpha\beta\gamma \cdots \perp\!\!\!\perp \lambda\mu\nu \cdots$ § 632.

For the converse theorem, since $\alpha\beta\gamma \cdots \perp\!\!\!\perp \lambda\mu\nu \cdots$ we hav $\alpha \perp\!\!\!\perp \lambda\mu\nu \cdots$, $\beta \perp\!\!\!\perp \lambda\mu\nu \cdots$, $\gamma \perp\!\!\!\perp \lambda\mu\nu \cdots$, $\cdots$ § 632.

Since $\alpha \perp\!\!\!\perp \lambda\mu\nu \cdots$, $\alpha \perp\!\!\!\perp \lambda$, $\alpha \perp\!\!\!\perp \mu$, $\alpha \perp\!\!\!\perp \nu$, $\cdots$

Similarly β, γ, $\cdots$ ar prime to λ, μ, ν, $\cdots$

634. Theorem. *If $\alpha \perp\!\!\!\perp \beta$, the product of any number of α's is prime to the product of any number of β's; and conversely, if the product of a number of α's is prime to the product of a number of β's, then $\alpha \perp\!\!\!\perp \beta$.*

This theorem is a particular case of the preceding

(1) See Dirichlet-Dedekind, p. 9.

theorem, the case when $\alpha, \beta, \gamma, \cdots$ ar the same integer and $\lambda, \mu, \nu, \cdots$ ar also the same.

635. Theorem. *If the product $\alpha\beta\gamma \cdots$ of two or more integers is divisible by a prime integer λ, one of the factors $\alpha, \beta, \gamma, \cdots$ must be divisible by λ* (1).

For, if none of the factors were divisible by λ, then $\lambda \neq 1$ and these factors would all be prime to λ.

§§ 443, 569.

Hence the product $\alpha\beta\gamma \cdots$ would be prime to λ.

§§ 557, 632.

Therfor this product would not be divisible by λ (§ 567), which contradicts the hypothesis.

636. Theorem. *If the products of two sets of prime positiv integers, no one of which is unity, ar equal, the two sets ar the same, except that the factors may be differently arranged in the two sets.*

Let $\alpha, \beta, \gamma, \cdots$ and $\lambda, \mu, \nu, \cdots$ be the two sets.

Then, since $\alpha\beta\gamma \cdots = \lambda\mu\nu \cdots$ and $\lambda\mu\nu \cdots$ is divisible by λ (§§ 381, 453), $\alpha\beta\gamma \cdots$ is divisible by λ. § 458.

Hence one of the integers $\alpha, \beta, \gamma, \cdots$, say α, is divisible by λ. § 635.

But α and λ ar primes and $\lambda \neq 1$.

Hence $\alpha = \lambda$. § 540.

Therfor $\beta\gamma \cdots = \mu\nu \cdots$ §§ 381, 341.

Similarly $\beta = \mu$.

And so forth.

Now ther must be the same number of integers in the two sets.

---

(1) See Tannery, p. 131.

For, suppose ther were not the same number.

By applying the above method we would, then, come to an equality of the form

$$\zeta = \rho\sigma\tau \cdots, \text{ or } \zeta . 1 = \rho\sigma\tau \cdots$$

By the same method as befor, we would now get $1 = \sigma\tau \cdots$

Hence 1 would be divisible by σ, an integer greater than 1, which is impossible.

Therfor the numbers of integers in the two sets must be the same, and the sets ar identical, except possibly for the order of their factors.

This theorem is usually briefly stated:

A positiv integer can be resolvd into positiv prime factors in only one way (1).

637. The representation of numbers by products of prime factors is a very useful artifice in many problems. Various operations ar renderd easier by this means and many properties of numbers ar made more evident (2).

We will giv a few illustrations.

638. Theorem. *If two or more positiv integers ar resolvd into prime factors, their product, resolvd into prime factors, will contain every distinct prime factor that is in either of the given numbers; if any prime factor occurs in only one of the given numbers, it will oc-*

(1) See K. F. Gauss, "Disquisitiones Arithmeticæ," 1801, p. 9; Chrystal, p. 44; Tannery, p. 132.

(2) See Tannery, p. 134.

cur the same number of times as a factor in the product; if any prime factor occurs in two or more of the given numbers, the number of times it will occur as a factor in the product will be the sum of the numbers of times that it occurs as a factor in these numbers ([1]).

Let the given positiv integers be $\nu_1, \nu_2, \nu_3, \cdots$

When resolvd into prime factors suppose

$$\nu_1=(\alpha\alpha\alpha\cdots a_1 \text{ times})(\beta\beta\beta\cdots b_1 \text{ times})(\gamma\gamma\gamma\cdots c_1 \text{ times})\cdots$$
$$\nu_2=(\alpha\alpha\alpha\cdots a_2 \text{ times})(\delta\delta\delta\cdots d_2 \text{ times})\cdots,$$
$$\nu_3=(\alpha\alpha\alpha\cdots a_3 \text{ times})(\gamma\gamma\gamma\cdots c_3 \text{ times})\cdots,$$
$$\cdots\cdots\cdots\cdots$$

The product of the given integers may then be written:

$$(\alpha\alpha\alpha\cdots a_1 \text{ times})(\beta\beta\beta\cdots b_1 \text{ times})(\gamma\gamma\gamma\cdots c_1 \text{ times})\cdots$$
$$\times\ (\alpha\alpha\alpha\cdots a_2 \text{ times})(\delta\delta\delta\cdots d_2 \text{ times})\cdots$$
$$\times\ (\alpha\alpha\alpha\cdots a_3 \text{ times})(\gamma\gamma\gamma\cdots c_3 \text{ times})\cdots$$
$$\times\ \cdots\cdots\cdots\cdots$$ § 384.

Since all the factors here written ar prime, this expression is the same as the expression for the product of the given numbers when it is resolvd into prime factors. § 636.

From its form it is evident that this expression contains every distinct prime factor that is in either of the given numbers.

Also that any prime factor, as δ, which occurs in only one of the given numbers occurs the same number of times as a factor in the product.

([1]) See Tannery, p. 134.

And that the number of times that any prime factor which is common to two or more of the given numbers, as α, occurs in the product is the sum of the numbers of times that it occurs in those numbers. § 37.

639. A convenient way of arranging the factors of the product given in the last article is to place together those that ar alike. We will then hav

$$\nu_1\nu_2\nu_3 \cdots = (\alpha\alpha\alpha \cdots (a_1 + a_2 + a_3) \text{ times})$$
$$\times (\beta\beta\beta \cdots b_1 \text{ times}) (\gamma\gamma\gamma \cdots (c_1 + c_3) \text{ times})$$
$$\times (\delta\delta\delta \cdots d_2 \text{ times}) \cdots$$

640. Theorem. *If two positiv integers α and β be resolvd into prime factors, then, if $\alpha \gtrdot \beta$, every prime factor that occurs in β will also occur in α and the number of times that it occurs in α will be not less than the number of times that it occurs in β; and conversely, if every prime factor that occurs in β occurs also in α a number of times not less than the number of times that it occurs in β, $\alpha \gtrdot \beta$* [1].

To prove the direct theorem, we hav $\alpha = \beta\gamma$.

The first part of the theorem is now evident from the first part of § 638.

Next let b be the number of times that any given prime factor occurs in β.

If this factor does not also occur in γ, it will occur in the product $\beta\gamma$, or in α, a number of times a, which is equal to b, by the second part of § 638.

If this factor occurs also in γ, say c times, it will

(1) See Legendre, p. 6; Stieltjes, p. 12.

occur in α a number of times a, which is equal to $b + c$, by the third part of § 638.

In this case $a > b$. § 75.

Hence in both cases $a \geqq b$.

Therfor $a \nless b$. § 129.

For the converse theorem, we may arrange the factors of α so that those which ar common to α and β shall come at the beginning, each the same number of times that it occurs in β. § 381.

We then hav $\alpha = \beta \times$ other factors.

Therfor $\alpha \mathrel{\dot{>}} \beta$.

For example, let $\alpha = \delta\delta\varepsilon\zeta\zeta\zeta$

and $\beta = \delta\zeta\zeta$.

We may then write $\alpha = (\delta\zeta\zeta)(\delta\epsilon\zeta) = \beta(\delta\varepsilon\zeta)$.

Hence $\alpha \mathrel{\dot{>}} \beta$.

641. Theorem. *If two positiv integers α and β be resolvd into prime factors, then, if $\alpha \mathrel{\dot{>}} \beta$, the quotient $\alpha : \beta$, resolvd into prime factors, will hav every prime factor that occurs in α but not in β the same number of times that it occurs in α; every prime factor that occurs the same number of times in α and β will not appear in the quotient; and every prime factor that occurs in both α and β but more times in α than in β will occur in the quotient a number of times equal to the difference of these numbers of times.*

We hav $\alpha = \beta\gamma$ and $\alpha : \beta = \gamma$.

Since the prime factors of the product $\beta\gamma$, or α, ar obtaind by putting together the prime factors of β

and γ, all the factors that α has but β has not must be factors of the quotient γ.

Hence the first part of the theorem is evident.

The second part of the theorem is true, for, if any prime factor occurd in both β and γ, it would occur in the product α a number of times equal to the sum of these numbers (§ 638), that is, the number of times it occurd in α would be greater than the number of times it occurd in β. § 75.

To prove the last part of the theorem, any factor that occurs in both α and β, but more times in α than in β must also occur in γ. Otherwise it would occur the same number of times in α as in β. § 638.

If, then, we let the numbers of these times be a, b, c respectivly, $a = b + c$. § 638.

Therfor $c = a - b$.

Thus, if $\alpha = \delta(\varepsilon\varepsilon\varepsilon \cdots e \text{ times})(\kappa\kappa\kappa \cdots k_1 \text{ times})$

and $\beta = (\varepsilon\varepsilon\varepsilon \cdots e \text{ times})(\kappa\kappa\kappa \cdots k_2 \text{ times})$,

$\alpha \geqslant \beta$ and $\alpha : \beta = \delta(\kappa\kappa\kappa \cdots (k_1 - k_2) \text{ times})$.

642. We will now study the theory of the greatest common factors of series of two or more integers and will find the following theorem useful.

Theorem. *The greatest common factor of two or more integers, of which at least one is not zero, is the same as the greatest common factor of those of the given integers which ar not zero.*

For every common factor of the integers which ar not zero is also a factor of the zero integers.

643. In virtue of this theorem, when treating of the

greatest common factor of a series of two or more integers, we will suppose that none of them is zero.

644. Theorem. *If a positiv integer* κ *is a common factor of a series of integers* $a_1, a_2, a_3, \cdots$ *and every positiv common factor of* $a_1, a_2, a_3, \cdots$ *is a factor of* κ, *then* κ *is the greatest common factor of* $a_1, a_2, a_3, \cdots$

Let λ be any positiv common factor of $a_1, a_2, a_3, \cdots$

Then by the hypothesis $\kappa = \varphi\lambda$ and $\lambda = \kappa : \varphi$, wher φ is positiv.

Now, if $\varphi > 1$, $\quad \kappa : \varphi < \kappa$, or $\lambda < \kappa$. § 493.

That is, of the positiv common factors the one for which $\varphi = 1$, which is κ, is larger than any other.

Therfor κ is the greatest common factor of $a_1, a_2, a_3, \cdots$

645. Theorem. *Having given any series of integers* $a_1, a_2, a_3, \cdots, a_n$, *none of which is zero, if we find the greatest common factor* κ_1 *of* a_1 *and* a_2, *then the greatest common factor* κ_2 *of* κ_1 *and* a_3, *then the greatest common factor* κ_3 *of* κ_2 *and* a_4, *and so on, and finally the greatest common factor* κ_{n-1} *of* κ_{n-2} *and* a_n, *then* κ_{n-1} *is the greatest common factor of* $a_1, a_2, a_3, \cdots, a_n$ (1).

First, to show that κ_{n-1} is a common factor of $a_1, a_2, a_3, \cdots, a_n$.

By hypothesis κ_1 is a common factor of a_1 and a_2.

Hence, since κ_2 is a factor of κ_1,
it is a common factor of a_1 and a_2. § 460.

And by hypothesis κ_2 is a factor of a_3.

---

(1) See Chrystal, p. 45; Tannery, p. 115; Le Besgue, p. 32; Lucas, p. 345.

Hence κ_2 is a common factor of α_1, α_2, α_3.

Since κ_3 is a factor of κ_2,
it is a common factor of α_1, α_2, α_3.

And by hypothesis κ_3 is a factor of α_4.

Hence κ_3 is a common factor of α_1, α_2, α_3, α_4.

Similarly κ_{n-1} is a common factor of α_1, α_2, α_3, $\cdots$, α_n.

Second, every positiv common factor of α_1, α_2, α_3, $\cdots$, α_n is a factor of κ_{n-1}.

For, let λ be any positiv common factor.

Then, since λ is a common factor of α_1 and α_2,
λ is a factor of κ_1. § 602.

Since λ is a common factor of κ_1 and α_3,
λ is a factor of κ_2.

Similarly we can prove that λ is a factor of κ_3, κ_4, $\cdots$, κ_{n-1}.

Since, then, κ_{n-1} is a common factor of α_1, α_2, α_3, $\cdots$, α_n and every positiv common factor of α_1, α_2, α_3, $\cdots$, α_n is a factor of κ_{n-1}, κ_{n-1} is the greatest common factor of α_1, α_2, α_3, $\cdots$, α_n. § 644.

This theorem might be stated :

Having given any series of integers α_1, α_2, α_3, $\cdots$, α_n, *none of which is zero, their greatest common factor is* $(((\alpha_1 \otimes \alpha_2) \otimes \alpha_3) \otimes \cdots) \otimes \alpha_n$.

646. Theorem. *If* α, β, γ *ar three integers, none of which is zero,* $(\alpha \otimes \beta) \otimes \gamma = \alpha \otimes (\beta \otimes \gamma)$.

By § 645 the first of these expressions is the greatest common factor of α, β, and γ.

So also is the second.

For $\alpha \otimes (\beta \otimes \gamma) = (\beta \otimes \gamma) \otimes \alpha$. § 574.

By § 645 $(\beta \otimes \gamma) \otimes \alpha$ is the greatest common factor of β, γ, and α.

But this is evidently the same as the greatest common factor of α, β, and γ.

Therfor $(\alpha \otimes \beta) \otimes \gamma = \alpha \otimes (\beta \otimes \gamma)$.

This theorem is the **Associativ Law for the Operation of finding Greatest Common Factor.**

The proof given above depends on the commutativ law. But it could easily be proved independently that $\alpha \otimes (\beta \otimes \gamma)$ is the greatest common factor of α, β, and γ. Then the associativ law would be independent of the commutativ law.

647. As a consequence of these two laws we can easily prove the following theorem, which is analogous to the generalized associativ and commutativ law for multiplication (§ 381), and which may be calld the **Generalized Associativ and Commutativ Law for the Operation of Finding Greatest Common Factor.**

Theorem. *In any complex expression in which integers, none of which is zero, ar connected by the sign for greatest common factor, with parentheses to indicate the order in which the various operations ar performd, the arrangement of the parentheses and the order of the elements may be changed in any way, the result of the whole set of operations being the greatest common factor of the series of integers.*

648. Hence we may write the expression $\alpha \otimes \beta \otimes \gamma \otimes \delta \otimes \varepsilon \otimes \cdots$ without parentheses, and may write the elements α, β, γ, δ, ε, $\cdots$ in any order we please, the expression still denoting the greatest common factor of α, β, γ, δ, ε, $\cdots$

649. Theorem. *If* $\alpha = \beta$, $\gamma = \delta$, $\varepsilon = \zeta$, $\eta = \theta$, $\cdots$, *then* $\alpha \otimes \gamma \otimes \varepsilon \otimes \eta \otimes \cdots = \beta \otimes \delta \otimes \zeta \otimes \theta \otimes \cdots$

This may be proved in the same way as § 384.

650. As a particular consequence of §§ 647, 649 it follows that in finding the greatest common factor of a series of integers, none of which is zero, we may replace any number of them by their greatest common factor and find the greatest common factor of this greatest common factor and the remaining integers (1).

651. Theorem. *If none of the integers* α_1, α_2, α_3, $\cdots$, α_n, θ *is zero,* $|\theta|(\alpha_1 \otimes \alpha_2 \otimes \alpha_3 \otimes \cdots \otimes \alpha_n)$

$$= (\theta\alpha_1) \otimes (\theta\alpha_2) \otimes (\theta\alpha_3) \otimes \cdots \otimes (\theta\alpha_n).$$

The proof of this theorem is formally the same as that of § 369.

This theorem is the **Generalized Left-handed Distributiv Law for Multiplication and the Operation of Finding Greatest Common Factor.**

652. Theorem. *If none of the integers* α_1, α_2, α_3, $\cdots$, α_n, θ *is zero,* $(\alpha_1 \otimes \alpha_2 \otimes \alpha_3 \otimes \cdots \otimes \alpha_n)|\theta|$

$$= (\alpha_1\theta) \otimes (\alpha_2\theta) \otimes (\alpha_3\theta) \otimes \cdots \otimes (\alpha_n\theta).$$

This theorem is the **Generalized Right-handed Distributiv Law for Multiplication and the Operation of Finding Greatest Common Factor.**

653. Theorem. *If the greatest common factor of a series of integers, none of which is zero, is unity, and these integers ar all multiplied by a given integer dif-*

(1) See Tannery, p. 115.

ferent from zero, the greatest common factor of the products is the numerical value of this multiplier; and conversely, if the greatest common factor of the products is the numerical value of the multiplier, the greatest common factor of the given integers is unity.

The proof of this theorem is like that of § 601.

654. Theorem. *If none of the integers*

$$\alpha_1, \alpha_2, \alpha_3, \cdots, \alpha_m, \beta_1, \beta_2, \beta_3, \cdots, \beta_n \text{ is zero,}$$

$$\begin{aligned}(\alpha_1 \otimes \alpha_2 \otimes \alpha_3 \otimes \cdots \otimes \alpha_n)(\beta_1 \otimes \beta_2 \otimes \beta_3 \otimes \cdots \otimes \beta_n) \\ = (\alpha_1\beta_1) \otimes (\alpha_2\beta_1) \otimes (\alpha_3\beta_1) \otimes \cdots \otimes (\alpha_m\beta_1) \\ \otimes (\alpha_1\beta_2) \otimes (\alpha_2\beta_2) \otimes (\alpha_3\beta_2) \otimes \cdots \otimes (\alpha_m\beta_2) \\ \otimes (\alpha_1\beta_3) \otimes (\alpha_2\beta_3) \otimes (\alpha_3\beta_3) \otimes \cdots \otimes (\alpha_m\beta_3) \\ \cdots\cdots\cdots\cdots\cdots\cdots \\ \otimes (\alpha_1\beta_n) \otimes (\alpha_2\beta_n) \otimes (\alpha_3\beta_n) \otimes \cdots \otimes (\alpha_m\beta_n)\end{aligned}$$ [1].

The proof of this theorem is like that of § 371.

This theorem may be stated: *The product of the respectiv greatest common factors of two series of integers, of which none is zero, is equal to the greatest common factor of all the products that can be formd by multiplying an integer of the first series by an integer of the second.*

655. Theorem. *The product of the respectiv greatest common factors of any number of series of integers, of which none is zero, is equal to the greatest common factor of all the products that can be formd, each having as a factor one and only one integer from each of the given series.*

---

[1] See Stieltjes, p. 3.

The proof of this theorem is formally the same as that of § 389.

656. Theorem. *If none of the integers* $a_1, a_2, a_3, \cdots$ *is zero and they ar all divisible by an integer* θ, *then* $a_1 \otimes a_2 \otimes a_3 \otimes \cdots \mathrel{\not<} \theta$ *and* $(a_1 \otimes a_2 \otimes a_3 \otimes \cdots) : |\theta| = (a_1 : \theta) \otimes (a_2 : \theta) \otimes (a_3 : \theta) \otimes \cdots$ (1).

This theorem is proved from § 652 just as § 602 is proved from § 599.

This theorem is the **Generalized Distributiv Law for Division and the Operation of Finding Greatest Common Factor.**

657. Theorem. *If all of a series of integers, none of which is zero, ar divided by their greatest common factor, the greatest common factor of the quotients will be unity; and conversely, if, when all of a series of integers, none of which is zero, ar divided by a given integer, the greatest common factor of the quotients is unity, the numerical value of the given integer is the greatest common factor of the given series of integers.*

The proof of this theorem is like that of § 603.

658. This theorem is equivalent to the following:

Theorem. *Having given a series of integers* $a_1, a_2, a_3, \ldots$; *if none of these integers is zero and* κ *is their greatest common factor, ther exists a series of integers* $\gamma_1, \gamma_2, \gamma_3, \cdots$, *none of which is zero and whose greatest common factor is unity, such that* $a_1 = \gamma_1\kappa$, $a_2 = \gamma_2\kappa$, $a_3 = \gamma_3\kappa, \cdots$; *and conversely, if* $a_1 = \gamma_1\kappa$, $a_2 = \gamma_2\kappa$, $a_3 = \gamma_3\kappa, \cdots$, *wher none of the integers* $\kappa, \gamma_1, \gamma_2, \gamma_3, \cdots$ *is zero and the greatest*

(1) See Stieltjes, p. 2.

common factor of $\gamma_1, \gamma_2, \gamma_3, \cdots$ *is unity, then* κ *is the greatest common factor of* $\alpha_1, \alpha_2, \alpha_3, \cdots$

659. Theorem. *The common factors of a series of integers, none of which is zero, ar the same as the factors of their greatest common factor* (1).

660. Theorem. *If two of the integers* $\alpha_1, \alpha_2, \alpha_3, \cdots$, *none of which is zero, ar prime to each other, the greatest common factor of the series is unity.*

To find the greatest common factor of the series of integers, we can arrange them as we please. § 647.

Call, then, the two that ar prime to each other α_1 and α_2.

Now, since $\alpha_1 \perp\!\!\!\perp \alpha_2$, $\alpha_1 \otimes \alpha_2 = 1$.

Then $\alpha_1 \otimes \alpha_2 \otimes \alpha_3 \otimes \cdots = (\alpha_1 \otimes \alpha_2) \otimes (\alpha_3 \otimes \cdots)$ § 647.

$= 1 \otimes (\alpha_3 \otimes \cdots)$ § 583.

$= 1.$ § 577.

661. Theorem. *If two of the integers* $\alpha_1, \alpha_2, \alpha_3, \cdots$, *none of which is zero, ar prime to each other, and* θ *is any integer except zero, then*

$$(\alpha_1\theta) \otimes (\alpha_2\theta) \otimes (\alpha_3\theta) \otimes \cdots = |\theta|.$$

This follows immediately from the preceding theorem and § 653.

662. Definition. A series of two or more integers ar **prime to each other two by two** (2), when each of these integers is prime to each of the others. Thus 2, 5, 9 ar prime to each other two by two.

(1) See Tannery, p. 112.

(2) See Chrystal, p. 38; Tannery, p. 117.

When two integers ar prime to each other, their greatest common factor is unity, and the converse is true (§ 579).

When three or more integers ar prime to each other two by two, their greatest common factor also is unity (§ 660), but the converse is not always true.

Thus the greatest common factor of 3, 5, 9 is unity, but these integers ar not prime to each other two by two, two pairs, 3 and 5, 5 and 9, being prime to each other, but the third pair 3 and 9 not being prime to each other.

Again the greatest common factor of 6, ϒ, ɜ̄ is unity, but no two of these integers ar prime to each other.

The statement that "When three or more integers ar prime to each other two by two, their greatest common factor is unity" is not, then, a complete analog of § 579.

We will, however, deduce in § 666 a theorem which is a complete analog of § 579, a theorem which is true of a series of integers and which includes that of § 579 as a particular case.

663. Theorem. *If* $\nu \underset{\cdot}{>} \theta a_1$, $\nu \underset{\cdot}{>} \theta a_2$, $\nu \underset{\cdot}{>} \theta a_3$, $\cdots$, $\nu \underset{\cdot}{>} \theta a_n$, *and the integers* $a_1, a_2, a_3, \cdots, a_n$ *ar prime to each other two by two, then* $\nu \underset{\cdot}{>} \theta a_1 a_2 a_3 \cdots a_n$ *and* $|\nu| : |\theta a_1 a_2 a_3 \cdots a_n| = (\nu : \theta a_1) \otimes (\nu : \theta a_2) \otimes (\nu : \theta a_3) \otimes \cdots \otimes (\nu : \theta a_n)$, *provided* ν *is not zero; and conversely, if* $\nu \underset{\cdot}{>} \theta a_1 a_2 a_3 \cdots a_n$, *then* $\nu \underset{\cdot}{>} \theta a_1$, $\nu \underset{\cdot}{>} \theta a_2$, $\nu \underset{\cdot}{>} \theta a_3$, $\cdots$, $\nu \underset{\cdot}{>} \theta a_n$, *and, if* $|\nu| : |\theta a_1 a_2 a_3 \cdots a_n|$

$= (\nu : \theta\alpha_1) \otimes (\nu : \theta\alpha_2) \otimes (\nu : \theta\alpha_3) \otimes \cdots \otimes (\nu : \theta\alpha_n)$, *then the integers* $\alpha_1, \alpha_2, \alpha_3, \cdots, \alpha_n$ *ar prime to each other two by two.*

For the direct theorem,

since $\nu \mathrel{\dot{>}} \theta\alpha_1$, $\nu \mathrel{\dot{>}} \theta\alpha_2$, and $\alpha_1 \perp\!\!\!\perp \alpha_2$, $\nu \mathrel{\dot{>}} \theta\alpha_1\alpha_2$

and $|\nu| : |\theta\alpha_1\alpha_2| = (\nu : \theta\alpha_1) \otimes (\nu : \theta\alpha_2)$. § 628.

That is, the theorem holds for two integers.

To prove it in general, suppose that it holds for k integers, wher $2 \leqq k < n$.

That is, we hav given $\nu \mathrel{\dot{>}} \theta\alpha_1\alpha_2\alpha_3 \cdots \alpha_k$ and

$|\nu| : |\theta\alpha_1\alpha_2\alpha_3 \cdots \alpha_k| = (\nu : \theta\alpha_1) \otimes (\nu : \theta\alpha_2) \otimes (\nu : \theta\alpha_3) \otimes \cdots \otimes (\nu : \theta\alpha_k)$.

Then $\alpha_1\alpha_2\alpha_3 \cdots \alpha_k \perp\!\!\!\perp \alpha_{k+1}$. § 632.

Hence $\nu \mathrel{\dot{>}} \theta(\alpha_1\alpha_2\alpha_3 \cdots \alpha_k)\alpha_{k+1}$ and

$|\nu| : |\theta(\alpha_1\alpha_2\alpha_3 \cdots \alpha_k)\alpha_{k+1}|$

$= (\nu : \theta(\alpha_1\alpha_2\alpha_3 \cdots \alpha_k)) \otimes (\nu : \theta\alpha_{k+1})$ § 628.

$= (|\nu| : |\theta\alpha_1\alpha_2\alpha_3 \cdots \alpha_k|) \otimes (\nu : \theta\alpha_{k+1})$ § 580.

$= (\nu : \theta\alpha_1) \otimes (\nu : \theta\alpha_2) \otimes (\nu : \theta\alpha_3) \otimes \cdots \otimes (\nu : \theta\alpha_k) \otimes (\nu : \theta\alpha_{k+1})$. § 583.

Hence, if the theorem holds for k integers, it holds also for $k + 1$.

Therfor by the law of induction it holds for n integers.

For the converse theorem we hav given

$$\nu \mathrel{\dot{>}} \theta\alpha_1\alpha_2\alpha_3 \cdots \alpha_n.$$

Hence $\nu \mathrel{\dot{>}} \theta\alpha_1$, $\nu \mathrel{\dot{>}} \theta\alpha_2$, $\nu \mathrel{\dot{>}} \theta\alpha_3$, $\cdots$, $\nu \mathrel{\dot{>}} \theta\alpha_n$. § 460.

Hence also $\nu = \lambda\theta\alpha_1\alpha_2\alpha_3 \cdots \alpha_n$, wher $\lambda \neq 0$.

Therfor $\nu : \theta\alpha_1\alpha_2\alpha_3 \cdots \alpha_n = \lambda$

and
$$\nu : \theta\alpha_1 = \lambda\alpha_2\alpha_3 \cdots \alpha_{n-1}\alpha_n,$$
$$\nu : \theta\alpha_2 = \lambda\alpha_1\alpha_3 \cdots \alpha_{n-1}\alpha_n,$$
$$\nu : \theta\alpha_3 = \lambda\alpha_1\alpha_2 \cdots \alpha_{n-1}\alpha_n,$$
$$\cdots\cdots\cdots\cdots,$$
$$\nu : \theta\alpha_n = \lambda\alpha_1\alpha_2\alpha_3 \cdots \alpha_{n-1}.$$

Hence
$$(\nu : \theta\alpha_1) \otimes (\nu : \theta\alpha_2) \otimes (\nu : \theta\alpha_3) \otimes \cdots \otimes (\nu : \theta\alpha_n)$$
$$= (\lambda\alpha_2\alpha_3 \cdots \alpha_{n-1}\alpha_n) \otimes (\lambda\alpha_1\alpha_3 \cdots \alpha_{n-1}\alpha_n)$$
$$\otimes (\lambda\alpha_1\alpha_2 \cdots \alpha_{n-1}\alpha_n) \otimes \cdots \otimes (\lambda\alpha_1\alpha_2\alpha_3 \cdots \alpha_{n-1}) \quad \text{§ 649.}$$
$$= |\lambda|[(\alpha_2\alpha_3 \cdots \alpha_{n-1}\alpha_n) \otimes (\alpha_1\alpha_3 \cdots \alpha_{n-1}\alpha_n)$$
$$\otimes (\alpha_1\alpha_2 \cdots \alpha_{n-1}\alpha_n) \otimes \cdots \otimes (\alpha_1\alpha_2\alpha_3 \cdots \alpha_{n-1})]. \quad \text{§ 651.}$$

But we hav given also
$$|\nu| : |\theta\alpha_1\alpha_2\alpha_3 \cdots \alpha_n|$$
$$= (\nu : \theta\alpha_1) \otimes (\nu : \theta\alpha_2) \otimes (\nu : \theta\alpha_3) \otimes \cdots \otimes (\nu : \theta\alpha_n).$$

Hence $|\lambda| = |\lambda|[(\alpha_2\alpha_3 \cdots \alpha_{n-1}\alpha_n) \otimes (\alpha_1\alpha_3 \cdots \alpha_{n-1}\alpha_n)$
$$\otimes (\alpha_1\alpha_2 \cdots \alpha_{n-1}\alpha_n) \otimes \cdots \otimes (\alpha_1\alpha_2\alpha_3 \cdots \alpha_{n-1})].$$

Therfor
$$(\alpha_2\alpha_3 \cdots \alpha_{n-1}\alpha_n) \otimes \{(\alpha_1\alpha_3 \cdots \alpha_{n-1}\alpha_n) \otimes (\alpha_1\alpha_2 \cdots \alpha_{n-1}\alpha_n)$$
$$\otimes \cdots \otimes (\alpha_1\alpha_2\alpha_3 \cdots \alpha_{n-1})\} = 1.$$

Or
$$(\alpha_2\alpha_3 \cdots \alpha_{n-1}\alpha_n) \otimes \{|\alpha_1|[(\alpha_3 \cdots \alpha_{n-1}\alpha_n) \otimes (\alpha_2 \cdots \alpha_{n-1}\alpha_n)$$
$$\otimes \cdots \otimes (\alpha_2\alpha_3 \cdots \alpha_{n-1})]\} = 1.$$

That is,
$$\alpha_2\alpha_3 \cdots \alpha_{n-1}\alpha_n \perp\!\!\perp |\alpha_1|[(\alpha_3 \cdots \alpha_{n-1}\alpha_n) \otimes (\alpha_2 \cdots \alpha_{n-1}\alpha_n)$$
$$\otimes \cdots \otimes (\alpha_2\alpha_3 \cdots \alpha_{n-1})]. \quad \text{§ 579.}$$

Therfor $\alpha_2\alpha_3 \cdots \alpha_{n-1}\alpha_n \perp\!\!\perp \alpha_1$. §§ 631, 561.

Therfor α_1 is prime to each of the integers $\alpha_2, \alpha_3, \cdots, \alpha_{n-1}, \alpha_n$. § 632.

Similarly it may be shown that each of the integers $\alpha_2, \alpha_3, \cdots, \alpha_n$ is prime to each of the other integers.

That is, $\alpha_1, \alpha_2, \alpha_3, \cdots, \alpha_n$ ar prime to each other two by two. § 662.

664. Theorem. *If* $\nu \mathrel{\dot{>}} \alpha_1, \quad \nu \mathrel{\dot{>}} \alpha_2, \quad \nu \mathrel{\dot{>}} \alpha_3, \cdots,$ *and the integers* $\alpha_1, \alpha_2, \alpha_3, \cdots$ *ar prime to each other two by two, then* $\nu \mathrel{\dot{>}} \alpha_1\alpha_2\alpha_3 \cdots$ *and*

$$|\nu| : |\alpha_1\alpha_2\alpha_3 \cdots| = (\nu : \alpha_1) \otimes (\nu : \alpha_2) \otimes (\nu : \alpha_3) \otimes \cdots,$$

provided ν *is not zero; and conversely, if* $\nu \mathrel{\dot{>}} \alpha_1\alpha_2\alpha_3 \cdots$, *then* $\nu \mathrel{\dot{>}} \alpha_1$, $\nu \mathrel{\dot{>}} \alpha_2$, $\nu \mathrel{\dot{>}} \alpha_3, \cdots$, *and, if*

$$|\nu| : |\alpha_1\alpha_2\alpha_3 \cdots| = (\nu : \alpha_1) \otimes (\nu : \alpha_2) \otimes (\nu : \alpha_3) \otimes \cdots,$$

then the integers $\alpha_1, \alpha_2, \alpha_3, \cdots$ *ar prime to each other two by two* (¹).

This theorem is a particular case of the preceding, the case when $\theta = 1$.

665. Theorem. *If* $\nu = \theta\alpha_1\alpha_2\alpha_3 \cdots$ *and the integers* $\alpha_1, \alpha_2, \alpha_3, \cdots$ *ar prime to each other two by two, then* $\nu \mathrel{\dot{>}} \theta\alpha_1, \quad \nu \mathrel{\dot{>}} \theta\alpha_2, \quad \nu \mathrel{\dot{>}} \theta\alpha_3, \cdots,$ *and* $(\nu : \theta\alpha_1) \otimes (\nu : \theta\alpha_2) \otimes (\nu : \theta\alpha_3) \otimes \cdots = 1$, *provided none of the integers* $\theta, \alpha_1, \alpha_2, \alpha_3, \cdots$ *is zero; and conversely, if* $(\nu : \theta\alpha_1) \otimes (\nu : \theta\alpha_2) \otimes (\nu : \theta\alpha_3) \otimes \cdots = 1$, *the integers* $\alpha_1, \alpha_2, \alpha_3, \cdots$ *ar prime to each other two by two.*

This follows immediately from § 663.

For, if $\nu = \theta\alpha_1\alpha_2\alpha_3 \cdots$, then $\nu \mathrel{\dot{>}} \theta\alpha_1$, $\nu \mathrel{\dot{>}} \theta\alpha_2$, $\nu \mathrel{\dot{>}} \theta\alpha_3, \cdots$, $\nu \mathrel{\dot{>}} \theta\alpha_1\alpha_2\alpha_3 \cdots$, and $\nu : \theta\alpha_1\alpha_2\alpha_3 \cdots = 1$.

666. Theorem. *If* $\nu = \alpha_1\alpha_2\alpha_3 \cdots$ *and the integers* $\alpha_1, \alpha_2, \alpha_3, \cdots$ *ar prime to each other two by two, then* $(\nu : \alpha_1) \otimes (\nu : \alpha_2) \otimes (\nu : \alpha_3) \otimes \cdots = 1$,

---

(¹) See Tannery, p. 120.

provided none of the integers $\alpha_1, \alpha_2, \alpha_3, \cdots$ *is zero; and conversely, if* $\nu = \alpha_1\alpha_2\alpha_3 \cdots$ *and* $(\nu : \alpha_1) \otimes (\nu : \alpha_2) \otimes (\nu : \alpha_3) \otimes \cdots = 1$, *the integers* $\alpha_1, \alpha_2, \alpha_3, \cdots$ *ar prime to each other two by two.*

This theorem might also be stated as follows:

If the integers $\alpha_1, \alpha_2, \alpha_3, \cdots, \alpha_n$, *none of which is zero, ar prime to each other two by two, the greatest common factor of the n products that can be formd, each containing all but one of these integers, is unity; and conversely, if the greatest common factor of these n products is unity, the given integers ar prime to each other two by two.*

For, if $\nu = \alpha_1\alpha_2\alpha_3 \cdots$, $\nu : \alpha_1 = \alpha_2\alpha_3 \cdots$,

$$\nu : \alpha_2 = \alpha_1\alpha_3 \cdots, \quad \nu : \alpha_3 = \alpha_1\alpha_2 \cdots, \quad \cdots$$

This theorem is the analog of § 579 mentiond in § 662.

It contains § 579 as a particular case.

For, if $\nu = \alpha_1\alpha_2$, $\nu : \alpha_1 = \alpha_2$ and $\nu : \alpha_2 = \alpha_1$.

Thus we hav from this theorem that, if $\alpha_1 \perp\!\!\perp \alpha_2$, $\alpha_2 \otimes \alpha_1 = 1$, and conversely.

The reason is now apparent why, in order that a series of n integers, when $n > 2$, may be prime to each other two by two, it is not sufficient that their greatest common factor shall be unity. The greatest common factor of the n products, each containing all but one of the integers, as above, must be unity.

But when $n = 2$, the n products ar the integers themselves, so that the two conditions ar the same.

667. Theorem. *If* κ *is the greatest common factor of* *the integers* $a_1, a_2, a_3, \cdots, a_n$, *none of which is zero, a series of integers* $\lambda_1, \lambda_2, \lambda_3, \cdots \lambda_n$ *can be found, such that*

$$\lambda_1 a_1 + \lambda_2 a_2 + \lambda_3 a_3 + \cdots + \lambda_n a_n = \kappa \text{ (1)}.$$

For, let κ_1 be the greatest common factor of a_1 and a_2, κ_2 of κ_1 and a_3, etc.

Then

$$\kappa_1 = \nu_1 a_1 + \xi_1 a_2,$$
$$\kappa_2 = \nu_2 \kappa_1 + \xi_2 a_3,$$
$$\kappa_3 = \nu_3 \kappa_2 + \xi_3 a_4,$$
$$\cdot \quad \cdot \quad \cdot \quad \cdot \quad \cdot \quad \cdot \quad \cdot,$$
$$\kappa_{n-1} = \nu_{n-1}\kappa_{n-2} + \xi_{n-1} a_n.$$

§ 610.

From the first two of these equalities

$$\kappa_2 = \nu_2(\nu_1 a_1 + \xi_1 a_2) + \xi_2 a_3 = \nu_2\nu_1 a_1 + \nu_2\xi_1 a_2 + \xi_2 a_3.$$

Combining this with the third, we get

$$\kappa_3 = \nu_3\nu_2\nu_1 a_1 + \nu_3\nu_2\xi_1 a_2 + \nu_3\xi_2 a_3 + \xi_3 a_4.$$

By examining this formula, we see that it is a particular case of the formula

$$\kappa_l = \nu_l\nu_{l-1}\cdots\nu_3\nu_2\nu_1 a_1 + \nu_l\nu_{l-1}\cdots\nu_3\nu_2\xi_1 a_2$$
$$+ \nu_l\nu_{l-1}\cdots\nu_3\xi_2 a_3 + \cdots + \nu_l\nu_{l-1}\xi_{l-2}a_{l-1} + \nu_l\xi_{l-1}a_l + \xi_l a_{l+1},$$

the case when $l = 3$.

Similarly the formulas for κ_1 and κ_2 ar particular cases of the latter formula.

We will now show that this formula holds for all values of l from 1 up to $n - 1$.

Suppose that it holds for some particular value l, wher $0 < l < n - 1$.

(1) See Gauss, p. 33.

Then, since $\kappa_{l+1} = \nu_{l+1}\kappa_l + \xi_{l+1}a_{l+2}$, we hav

$$\kappa_{l+1} = \nu_{l+1}\nu_l\nu_{l-1}\cdots\nu_3\nu_2\nu_1a_1 + \nu_{l+1}\nu_l\nu_{l-1}\cdots\nu_3\nu_2\xi_1a_2$$
$$+ \nu_{l+1}\nu_l\nu_{l-1}\cdots\nu_3\xi_2a_3 + \cdots + \nu_{l+1}\nu_l\nu_{l-1}\xi_{l-2}a_{l-1}$$
$$+ \nu_{l+1}\nu_l\xi_{l-1}a_l + \nu_{l+1}\xi_la_{l+1} + \xi_{l+1}a_{l+2}.$$

But this is the same as the formula for κ_l with $l+1$ put in place of l.

If, then, the formula for κ_l holds for any particular value of l, it holds for the next.

Hence, since we hav shown that it holds for the values 1, 2, 3, by the law of mathematical induction it holds for all values of l from 1 up to the last value $n-1$.

We hav, then, finally

$$\kappa_{n-1} = \nu_{n-1}\nu_{n-2}\cdots\nu_3\nu_2\nu_1a_1 + \nu_{n-1}\nu_{n-2}\cdots\nu_3\nu_2\xi_1a_2$$
$$+ \nu_{n-1}\nu_{n-2}\cdots\nu_3\xi_2a_3 + \cdots + \nu_{n-1}\nu_{n-2}\xi_{n-3}a_{n-2}$$
$$+ \nu_{n-1}\xi_{n-2}a_{n-1} + \xi_{n-1}a_n.$$

But $\kappa_{n-1} = \kappa$.

Hence, setting
$$\lambda_1 = \nu_{n-1}\nu_{n-2}\cdots\nu_3\nu_2\nu_1,$$
$$\lambda_2 = \nu_{n-1}\nu_{n-2}\cdots\nu_3\nu_2\xi_1,$$
$$\lambda_3 = \nu_{n-1}\nu_{n-2}\cdots\nu_3\xi_2,$$
$$\cdots\cdots\cdots,$$

we hav $\kappa = \lambda_1a_1 + \lambda_2a_2 + \lambda_3a_3 + \cdots + \lambda_na_n$.

668. Theorem. *If unity is the greatest common factor of a series of integers $a_1, a_2, a_3, \cdots, a_n$, none of which is zero, another series of integers $\lambda_1, \lambda_2, \lambda_3, \cdots, \lambda_n$ can be found, such that*

$$\lambda_1a_1 + \lambda_2a_2 + \lambda_3a_3 + \cdots + \lambda_na_n = 1.$$

669. Theorem. *If* $\alpha_1 + \alpha_2 + \alpha_3 + \cdots = 1$, *then* $\alpha_1 \otimes \alpha_2 \otimes \alpha_3 \otimes \cdots = 1$.

This is proved in the same way as § 612.

670. Theorem. *If* $\alpha_1 + \alpha_2 + \alpha_3 + \cdots = 1$ *and* β_1, β_2, β_3, $\cdots$ *ar factors of* α_1, α_2, α_3, $\cdots$ *respectivly, then* $\beta_1 \otimes \beta_2 \otimes \beta_3 \otimes \cdots = 1$.

671. Theorem. *If* $\lambda_1\alpha_1 + \lambda_2\alpha_2 + \lambda_3\alpha_3 + \cdots = 1$, *the greatest common factor of any series of integers taken one from each of the groups* λ_1 *and* α_1, λ_2 *and* α_2, λ_3 *and* α_3, $\cdots$ *is unity.*

672. Theorem.

If $\lambda_1\alpha_1 + \lambda_2\alpha_2 + \lambda_3\alpha_3 + \cdots = \alpha_1 \otimes \alpha_2 \otimes \alpha_3 \otimes \cdots$,
then $\lambda_1 \otimes \lambda_2 \otimes \lambda_3 \otimes \cdots = 1$.

This is proved in the same way as § 616.

This theorem may also be stated: *In whatever may the greatest common factor of* α_1, α_2, α_3, $\cdots$ *is exprest in the form* $\lambda_1\alpha_1 + \lambda_2\alpha_2 + \lambda_3\alpha_3 + \cdots$, *the greatest common factor of* λ_1, λ_2, λ_3, $\cdots$ *is unity.*

673. Theorem. *If* κ *is the positiv integer which has as prime factors all the prime factors that ar common to a series of positiv integers, and which contains each prime factor a number of times equal to the least number of times that it occurs as a factor in the given integers, then* κ *is the greatest common factor of those integers.*

First, since each of the given integers contains every prime factor that occurs in κ a number of times not less than the number of times that it occurs in κ, κ is a common factor of the given integers. § 640.

Now let λ be any positiv common factor of the given integers.

Then every prime factor of λ must be common to the given integers and must occur in λ a number of times not greater than the number of times that it occurs in any one of the given integers (§ 640), and hence not greater than the least number of times that it occurs as a factor in the given integers.

Therfor every prime factor of λ occurs in κ a number of times not less than the number of times that it occurs in λ.

Hence λ is a factor of κ.

Since, therfor, κ is a common factor of the given integers and every positiv common factor is a factor of κ, κ is the greatest common factor of the given integers.

CHAPTER IX.

LEAST COMMON MULTIPLE.

674. Definition. A **common multiple** [1] of two, or more, integers is an integer that is a multiple of each of them.

Thus zero is a common multiple of any two, or more, integers, none of which is zero; the product of two, or more, integers, none of which is zero, is a multiple of each of them; also their product multiplied by any factor.

675. Let $\alpha, \beta, \gamma, \cdots$ be any two, or more, integers, neither of which is zero, and λ any integer not zero. Then $\lambda\alpha\beta\gamma\cdots$ is a common multiple of $\alpha, \beta, \gamma, \cdots$ Now of all the values of $\lambda\alpha\beta\gamma\cdots$ obtaind by giving λ various values the numerically smallest ar those for which $|\lambda| = 1$. For,

if $\quad |\lambda| > 1, \qquad |\lambda\alpha\beta\gamma\cdots| > |\alpha\beta\gamma\cdots|. \qquad$ § 406.

Ther may be common multiples of $\alpha, \beta, \gamma, \cdots$ which ar numerically greater than $\alpha\beta\gamma\cdots$, but which ar not multiples of $\alpha\beta\gamma\cdots$ Thus 9×4, or 6×6, which is greater than 4×6, is a common multiple of 4 and 6, but is not a multiple of 4×6.

Moreover ther may be common multiples of $\alpha, \beta,$

(1) See Stieltjes, p. 2; Tannery, p. 121.

$\gamma, \cdots$ numerically less than the product $\alpha\beta\gamma\cdots$ Thus 3×4, or 2×6, which is a common multiple of 4 and 6, is less than the product 4×6.

676. The number of positiv integers not greater than $|\alpha\beta\gamma\cdots|$ is $|\alpha\beta\gamma\cdots|$. Hence the number of positiv common multiples of $\alpha, \beta, \gamma, \cdots$ not greater than $|\alpha\beta\gamma\cdots|$ is not greater than $|\alpha\beta\gamma\cdots|$. It is, therfor, limited. Of these multiples any one is less than any positiv common multiple greater than $|\alpha\beta\gamma\cdots|$. The smallest of them is therfor the smallest of all the positiv common multiples of $\alpha, \beta, \gamma, \cdots$ It is calld the **least common multiple** of $\alpha, \beta, \gamma, \cdots$

Definition. The **least common multiple** of two, or more, integers, none of which is zero, is the smallest of the positiv common multiples of those integers.

677. If we had defined the least common multiple of a set of integers as the numerical value of the numerically smallest of their multiples, the least common multiple of any set of integers would be zero.

The idea of least common multiple would not, then, be anything new. The idea as defined above, however, is a new and useful idea. The least common multiples of different sets of integers ar not always the same, as may be seen by taking a few examples.

It is to be noted also that the least common multiple is not defined except for a set of integers, none of which is zero. This is in accordance with the restriction of § 427.

678. Theorem. *If a positiv integer μ is a common*

multiple of a series of integers $\alpha_1, \alpha_2, \alpha_3, \cdots$ *and every positiv common multiple of* $\alpha_1, \alpha_2, \alpha_3, \cdots$ *is a multiple of* μ, *then* μ *is the least common multiple of* $\alpha_1, \alpha_2, \alpha_3, \cdots$

Let ν be any positiv common multiple of $\alpha_1, \alpha_2, \alpha_3, \cdots$

Then by the hypothesis $\nu = \varphi\mu$, wher φ is positiv.

Now, if $\varphi > 1$, $\varphi\mu > \mu$, or $\nu > \mu$.

That is, of the positiv common multiples the one for which $\varphi = 1$, which is μ, is less than any other.

Therfor μ is the least common multiple of $\alpha_1, \alpha_2, \alpha_3, \cdots$

679. We will denote the least common multiple of two integers α and β by the symbol $\alpha \times \beta$, which may be read "α mul β." The symbol $\alpha \times \beta$ will, then, be univalent, if neither α nor β is zero. The symbols $\alpha \times 0$, $0 \times \alpha$, and 0×0 will, however, be meaningless.

The integers α and β may be calld the **elements** of the least common multiple $\alpha \times \beta$.

680. Operation of Finding Least Common Multiple. The least common multiple of two integers, neither of which is zero, is found by choosing the smallest among a given limited set of integers. It is therfor found by an operation.

681. Theorem. $\alpha \times \beta = \beta \times \alpha$, *provided neither* α *nor* β *is zero.*

The proof is like that of § 574.

This theorem is the **Commutativ Law for the Operation of Finding Least Common Multiple.**

682. Theorem. *If* $\alpha \mathrel{\dot{>}} \beta$, $\alpha \times \beta = |\alpha|$, *provided neither* α *nor* β *is zero.*

The proof is like that of § 575.

683. Theorem. $\alpha \underline{\times} \alpha = |\alpha|$, *provided* α *is not zero.*

684. Theorem. $\alpha \underline{\times} 1 = |\alpha|$, *provided* α *is not zero.*

685. Theorem.

$$(-\alpha) \underline{\times} \beta = \alpha \underline{\times} (-\beta) = (-\alpha) \underline{\times} (-\beta) = \alpha \underline{\times} \beta,$$

provided neither α *nor* β *is zero.*

686. Theorem. *If* $\alpha = \beta$, $\alpha \underline{\times} \gamma = \beta \underline{\times} \gamma$, *provided neither* β *nor* γ *is zero.*

687. Theorem. *If* $\alpha = \beta$, $\gamma \underline{\times} \alpha = \gamma \underline{\times} \beta$, *provided neither* β *nor* γ *is zero.*

688. Theorem. *If* $\alpha = \beta$ *and* $\gamma = \delta$, $\alpha \underline{\times} \gamma = \beta \underline{\times} \delta$, *provided neither* β *nor* γ *is zero.*

689. As the least common multiple of two integers is an integer, we may combine it with other integers by any of the signs $+$, $-$, $\times$, $:$, $\otimes$, $\underline{\times}$; provided, first, that the first element of every quotient is divisible by the second; second, that of two integers connected by the sign $\otimes$ not both ar zero; and third, that of two integers connected by the sign $\underline{\times}$ neither is zero. We will use parentheses to indicate the order in which the various operations ar performd.

Every such expression is univalent. For every sum, difference, product, quotient, greatest common factor, and least common multiple is univalent.

690. Theorem. *If for one or more elements of a complex expression which contains integers connected by any or all of the signs* $+$, $-$, $\times$, $:$, $\otimes$, $\underline{\times}$ *and in which, first, the first element of every quotient is divisible by the second element, second, of two integers connected by* *the sign* $\otimes$ *not both ar zero, and third, of two integers connected by the sign* $\times$ *neither is zero, equal integers ar substituted, the complex expression is unchanged.*

691. Theorem. $\alpha \times \beta = |\alpha\beta| : (\alpha \otimes \beta)$, *provided neither* α *nor* β *is zero* (1).

Since neither α nor β is zero, $|\alpha\beta|$ is positiv.

Hence $|\alpha\beta| : (\alpha \otimes \beta)$ is positiv.

Moreover $|\alpha\beta| : (\alpha \otimes \beta) = |\alpha| (|\beta| : (\alpha \otimes \beta))$ § 468.

$= (|\alpha| : (\alpha \otimes \beta))|\beta|$. § 467.

Therfor $|\alpha\beta| : (\alpha \otimes \beta)$ is a common multiple of $|\alpha|$ and $|\beta|$, and hence of α and β.

Moreover every common multiple of α and β is a multiple of $|\alpha\beta| : (\alpha \otimes \beta)$. § 630.

Therfor $|\alpha\beta| : (\alpha \otimes \beta)$ is the least common multiple of α and β. § 678.

By means of this theorem the least common multiple of two integers, neither of which is zero, may be easily found when their product and greatest common factor ar given (2).

692. Theorem. *Any common multiple* ν *of two integers* α *and* β *is a multiple of their least common multiple.*

If ν is zero, the theorem follows from § 444.

If ν is not zero, $\nu \geqslant |\alpha\beta| : (\alpha \otimes \beta)$. § 630.

And $|\alpha\beta| : (\alpha \otimes \beta) = \alpha \times \beta$. § 691.

Hence $\nu \geqslant \alpha \times \beta$.

---

(1) See Le Besgue, p. 34; Tannery, p. 122.

(2) See Stieltjes, p. 4.

693. Theorem. *The common multiples of two integers ar the same as the multiples of their least common multiple* [1].

For, by § 460, every multiple of the least common multiple is a multiple of each of the given integers.

And, by § 692, every common multiple of the given integers is a multiple of their least common multiple.

Hence the two groups of multiples, first the common multiples of the two integers and second the multiples of their least common multiple ar the same.

694. Theorem. $(\alpha \times \beta)(\alpha \otimes \beta) = |\alpha\beta|$, *provided neither* α *nor* β *is zero.*

This follows easily from § 691.

As a consequence of this theorem ther is a **reciprocity** or **duality** [2] between the theories of greatest common factor and least common multiple. It will be found that many theorems in these subjects ar still true when the terms "greatest common factor" and "least common multiple" ar interchanged.

695. Theorem. $\alpha \otimes \beta = |\alpha\beta| : (\alpha \times \beta)$, *provided neither* α *nor* β *is zero.*

By means of this theorem the greatest common factor of two integers, neither of which is zero, may

(1) See Tannery, p. 122.

(2) See L. Cremona, "Eléments de Géométrie Projective," 1872, English translation, "Elements of Projective Geometry," 1885, by C. Leudesdorf, p. 26; Beman and Smith, "New Plane and Solid Geometry," 1901, p. 27.

be easily found when their product and least common multiple ar given.

696. Theorem.

If ν is a common multiple of α and β, then

$$[(\nu : \alpha) \times (\nu : \beta)]\,[(\nu : \alpha) \otimes (\nu : \beta)] = |\nu\nu| : |\alpha\beta|,$$

provided ν is not zero.

For $[(\nu : \alpha) \times (\nu : \beta)]\,[(\nu : \alpha) \otimes (\nu : \beta)]$

$= |(\nu : \alpha)(\nu : \beta)|$ § 694.

$= |(\nu\nu) : (\alpha\beta)|$ § 466.

$= |\nu\nu| : |\alpha\beta|.$ § 496.

697. Theorem.

If ν is a common multiple of α and β, then

$(\alpha \times \beta)[(\nu : \alpha) \otimes (\nu : \beta)] = |\nu|$, *provided ν is not zero.*

By § 630, $|\nu| : (|\alpha\beta| : (\alpha \otimes \beta)) = (\nu : \alpha) \otimes (\nu : \beta)$.

But $|\alpha\beta| : (\alpha \otimes \beta) = \alpha \times \beta$. § 691.

Therfor $|\nu| : (\alpha \times \beta) = (\nu : \alpha) \otimes (\nu : \beta)$.

Or $(\alpha \times \beta)[(\nu : \alpha) \otimes (\nu : \beta)] = |\nu|$.

This theorem includes § 694 as a particular case.

For, if $\nu = \alpha\beta$, $\nu : \alpha = \beta$ and $\nu : \beta = \alpha$.

698. Theorem. *If ν is the least common multiple of the integers α and β, then $(\nu : \alpha) \otimes (\nu : \beta) = 1$; and conversely, if the positiv integer ν is a common multiple of the integers α and β and $(\nu : \alpha) \otimes (\nu : \beta) = 1$, then ν is the least common multiple of α and β.*

Both parts of the theorem follow easily from § 697.

699. Theorem. *If ν is a common multiple of α and β, then $(\alpha \otimes \beta)\,[(\nu : \alpha) \times (\nu : \beta)] = |\nu|$, provided ν is not zero.*

For

$(\alpha \times \beta)\{(\alpha \otimes \beta)[(\nu : \alpha) \times (\nu : \beta)]\}[(\nu : \alpha) \otimes (\nu : \beta)]$
$= \{(\alpha \times \beta)(\alpha \otimes \beta)\}$
$\quad \times \{[(\nu : \alpha) \times (\nu : \beta)][(\nu : \alpha) \otimes (\nu : \beta)]\}$ § 381.
$= |\alpha\beta| \times (|\nu\nu| : |\alpha\beta|) = |\nu\nu|.$ §§ 694, 696.

Moreover

$(\alpha \times \beta)\{(\alpha \otimes \beta)[(\nu : \alpha) \times (\nu : \beta)]\}[(\nu : \alpha) \otimes (\nu : \beta)]$
$= \{(\alpha \otimes \beta)[(\nu : \alpha) \times (\nu : \beta)]\}$
$\quad \times \{(\alpha \times \beta)[(\nu : \alpha) \otimes (\nu : \beta)]\}$ § 381.
$= \{(\alpha \otimes \beta)[(\nu : \alpha) \times (\nu : \beta)]\} |\nu|.$ § 697.

Hence $\{(\alpha \otimes \beta)[(\nu : \alpha) \times (\nu : \beta)]\} |\nu| = |\nu\nu|.$
Therfor $(\alpha \otimes \beta)[(\nu : \alpha) \times (\nu : \beta)] = |\nu|.$

700. Theorem. *If* $\alpha \perp\!\!\perp \beta$, $\alpha \times \beta = |\alpha\beta|$, *provided neither* α *nor* β *is zero, and conversely.*

This follows easily from §§ 579, 691.

701. Theorem. $(\alpha\beta) \times (\alpha\gamma) = |\alpha|(\beta \times \gamma)$, *provided neither* α, β, *nor* γ *is zero.*

For $(\alpha\beta) \times (\alpha\gamma) = |(\alpha\beta)(\alpha\gamma)| : ((\alpha\beta) \otimes (\alpha\gamma))$ § 691.
$= (|\alpha||\alpha||\beta\gamma|) : (|\alpha|(\beta \otimes \gamma))$
$= (|\alpha||\beta\gamma|) : (\beta \otimes \gamma)$ § 457.
$= |\alpha|(|\beta\gamma| : (\beta \otimes \gamma)) = |\alpha|(\beta \times \gamma).$ §§ 468, 691.

This theorem is the **Left-handed Distributiv Law for Multiplication and the Operation of Finding Least Common Multiple.**

702. Theorem. $(\alpha\gamma) \times (\beta\gamma) = (\alpha \times \beta)|\gamma|$, *provided neither* α, β, *nor* γ *is zero.*

This theorem is the **Right-handed Distributiv Law for Multiplication and the Operation of Finding Least Common Multiple.**

703. Theorem. *If* $\beta \amalg \gamma$, $(\alpha\beta) \times (\alpha\gamma) = |\,\alpha\beta\gamma\,|$, *provided neither* α, β, *nor* γ *is zero.*

For $(\alpha\beta) \times (\alpha\gamma) = |\,\alpha\,|\,(\beta \times \gamma)$ § 701.

$= |\,\alpha\,|\,|\,\beta\gamma\,| = |\,\alpha\beta\gamma\,|$. § 700.

704. Theorem. *If* $\alpha \dot{>} \gamma$ *and* $\beta \dot{>} \gamma$, *then* $\alpha \times \beta \dot{>} \gamma$ *and* $(\alpha \times \beta) : |\,\gamma\,| = (\alpha : \gamma) \times (\beta : \gamma)$, *provided neither* α *nor* β *is zero.*

This theorem may be proved in the same way as § 602, or it may be proved from § 602 by means of § 691.

This theorem is the **Distributiv Law for Division and the Operation of Finding Least Common Multiple.**

705. Theorem. *If* $\theta \amalg \beta$, $(\alpha\theta) \times \beta = (\alpha \times \beta)\,|\,\theta\,|$, *provided neither* α, β, *nor* θ *is zero.*

This may be proved from § 625 by means of § 691.

706. Theorem. *If* $\alpha \dot{>} \theta$ *and* $\theta \amalg \beta$, *then* $(\alpha : \theta) \times \beta = (\alpha \times \beta) : |\,\theta\,|$, *provided neither* α *nor* β *is zero.*

This may be proved from § 626 by means of § 691.

707. Theorem. *Having given any series of integers* $\alpha_1, \alpha_2, \alpha_3, \cdots, \alpha_n$, *none of which is zero, if we find the least common multiple* μ_1 *of* α_1 *and* α_2, *then the least common multiple* μ_2 *of* μ_1 *and* α_3, *then the least common multiple* μ_3 *of* μ_2 *and* α_4, *and so on, and finally the least common multiple* μ_{n-1} *of* μ_{n-2} *and* α_n, *then* μ_{n-1} *is the least common multiple of* $\alpha_1, \alpha_2, \alpha_3, \cdots, \alpha_n$ (1).

First, to show that μ_{n-1} is a common multiple of $\alpha_1, \alpha_2, \alpha_3, \cdots, \alpha_n$.

(1) See Tannery, p. 122; Lucas, p. 345.

By hypothesis μ_1 is a common multiple of a_1 and a_2.

Hence, since μ_2 is a multiple of μ_1,
it is a common multiple of a_1 and a_2. § 460.

And by hypothesis μ_2 is a multiple of a_3.

Hence μ_2 is a common multiple of a_1, a_2, a_3.

Since μ_3 is a multiple of μ_2,
it is a common multiple of a_1, a_2, a_3.

And by hypothesis μ_3 is a multiple of a_4.

Hence μ_3 is a common multiple of a_1, a_2, a_3, a_4.

Similarly μ_{n-1} is a common multiple of a_1, a_2, a_3, $\cdots$, a_n.

Second, every positiv common multiple of a_1, a_2, a_3, $\cdots$, a_n is a multiple of μ_{n-1}.

For, let ν be any positiv common multiple.

Then, since ν is a common multiple of a_1 and a_2,
ν is a multiple of μ_1. § 692.

Since ν is a common multiple of μ_1 and a_3,
ν is a multiple of μ_2.

Similarly we may prove that ν is a multiple of μ_3, μ_4, $\cdots$, μ_{n-1}.

Since, then, μ_{n-1} is a common multiple of a_1, a_2, a_3, $\cdots$, a_n and every positiv common multiple of a_1, a_2, a_3, $\cdots$, a_n is a multiple of μ_{n-1}, μ_{n-1} is the least common multiple of a_1, a_2, a_3, $\cdots$, a_n. § 678.

This theorem might be stated:

Having given any series of integers a_1, a_2, a_3, $\cdots$, a_n, *none of which is zero, their least common multiple is* $(((a_1 \times a_2) \times a_3) \times \cdots) \times a_n$.

It will be noticed that this theorem and its proof ar identically the same as those of § 645, except that the words "least" and "multiple" replace "greatest" and "factor" respectivly, the letter μ replaces κ, and the symbol $\times$ replaces $\otimes$.

708. Theorem. *If* α, β, γ *ar three integers, none of which is zero*, $(\alpha \times \beta) \times \gamma = \alpha \times (\beta \times \gamma)$.

The proof is similar in form to that of § 646.

This theorem is the **Associativ Law for the Operation of Finding Least Common Multiple.**

709. As a consequence of this theorem and that of § 681 we can prove the following theorem, which is analogous to that of § 647 and which may be calld the **Generalized Associativ and Commutativ Law for the Operation of Finding Least Common Multiple.**

Theorem. *In any complex expression in which integers, none of which is zero, ar connected by the sign for least common multiple, with parentheses to indicate the order in which the various operations ar performd, the arrangement of the parentheses and the order of the elements may be changed in any way, the result of the whole set of operations being the least common multiple of the series of integers* (1).

710. Hence we may write the expression $\alpha \times \beta \times \gamma \times \delta \times \varepsilon \times \cdots$ without parentheses, and may write the elements α, β, γ, δ, ε, $\cdots$ in any order we please, the expression still denoting the least common multiple of α, β, γ, δ, ε, $\cdots$

(1) See Le Besgue, p. 34.

711. Theorem. *If* $\alpha = \beta$, $\gamma = \delta$, $\varepsilon = \zeta$, $\eta = \theta$, $\cdots$, *then* $\alpha \times \gamma \times \varepsilon \times \eta \times \cdots = \beta \times \delta \times \zeta \times \theta \cdots$

This may be proved in the same way as § 649.

712. As a particular consequence of §§ 709, 711 it follows that in finding the least common multiple of a series of integers, none of which is zero, we may replace any number of them by their least common multiple and find the least common multiple of this least common multiple and the remaining integers (1).

713. Theorem. *If none of the integers* $\alpha_1, \alpha_2, \alpha_3, \cdots, \alpha_n, \theta$ *is zero,*

$$|\theta|(\alpha_1 \times \alpha_2 \times \alpha_3 \times \cdots \times \alpha_n) = (\theta\alpha_1) \times (\theta\alpha_2) \times (\theta\alpha_3) \times \cdots \times (\theta\alpha_n)$$ (2).

The proof of this theorem is formally the same as that of § 651.

This theorem is the **Generalized Left-handed Distributiv Law for Multiplication and the Operation of Finding Least Common Multiple.**

714. Theorem. *If none of the integers* $\alpha_1, \alpha_2, \alpha_3, \cdots, \alpha_n, \theta$ *is zero,*

$$(\alpha_1 \times \alpha_2 \times \alpha_3 \times \cdots \times \alpha_n)|\theta| = (\alpha_1\theta) \times (\alpha_2\theta) \times (\alpha_3\theta) \times \cdots \times (\alpha_n\theta).$$

This theorem is the **Generalized Right-handed Distributiv Law for Multiplication and the Operation of Finding Least Common Multiple.**

715. Theorem. *If none of the integers* $\alpha_1, \alpha_2, \alpha_3, \cdots, \alpha_m, \beta_1, \beta_2, \beta_3, \cdots, \beta_n$ *is zero,*

(1) See Tannery, p. 123. (2) See Tannery, p. 124.

$$(\alpha_1 \underline{\times} \alpha_2 \underline{\times} \alpha_3 \underline{\times} \cdots \underline{\times} \alpha_m)(\beta_1 \underline{\times} \beta_2 \underline{\times} \beta_3 \underline{\times} \cdots \underline{\times} \beta_n)$$
$$= (\alpha_1\beta_1) \underline{\times} (\alpha_2\beta_1) \underline{\times} (\alpha_3\beta_1) \underline{\times} \cdots \underline{\times} (\alpha_m\beta_1)$$
$$\underline{\times} (\alpha_1\beta_2) \underline{\times} (\alpha_2\beta_2) \underline{\times} (\alpha_3\beta_2) \underline{\times} \cdots \underline{\times} (\alpha_m\beta_2)$$
$$\underline{\times} (\alpha_1\beta_3) \underline{\times} (\alpha_2\beta_3) \underline{\times} (\alpha_3\beta_3) \underline{\times} \cdots \underline{\times} (\alpha_m\beta_3)$$
$$\underline{\times} \cdot \cdot \cdot \cdot \cdot \cdot \cdot \cdot \cdot \cdot \cdot \cdot \cdot \cdot \cdot \cdot \cdot \cdot$$
$$\underline{\times} (\alpha_1\beta_n) \underline{\times} (\alpha_2\beta_n) \underline{\times} (\alpha_3\beta_n) \underline{\times} \cdots \underline{\times} (\alpha_m\beta_n)$$ (1).

The proof of this theorem is like that of § 654.

This theorem may be stated: *The product of the respectiv least common multiples of two series of integers, of which none is zero, is equal to the least common multiple of all the products that can be formd by multiplying an integer of the first series by an integer of the second.*

716. Theorem. *The product of the respectiv least common multiples of any number of series of integers, of which none is zero, is equal to the least common multiple of all the products that can be formd each having as a factor one and only one integer from each of the given series.*

The proof of this theorem is formally the same as that of § 655.

717. Theorem. *If none of the integers $\alpha_1, \alpha_2, \alpha_3, \cdots$ is zero and they ar all divisible by an integer θ, then $\alpha_1 \underline{\times} \alpha_2 \underline{\times} \alpha_3 \underline{\times} \cdots \gtrdot \theta$ and*

$$(\alpha_1 \underline{\times} \alpha_2 \underline{\times} \alpha_3 \underline{\times} \cdots) : |\theta|$$
$$= (\alpha_1 : \theta) \underline{\times} (\alpha_2 : \theta) \underline{\times} (\alpha_3 : \theta) \underline{\times} \cdots$$

The proof of this theorem is formally the same as that of § 656.

(1) See Stieltjes, p. 3.

This theorem is the **Generalized Distributiv Law for Division and the Operation of Finding Least Common Multiple.**

718. Theorem. *The common multiples of a series of integers $a_1, a_2, a_3, \cdots, a_n$ ar the same as the multiples of their least common multiple* (1).

For, let $\mu_1, \mu_2, \mu_3, \cdots, \mu_{n-1}$ be the series of least common multiples found as in § 707.

Then the common multiples of a_1 and a_2 ar the same as the multiples of μ_1. § 693.

The common multiples of μ_1 and a_3 ar the same as the multiples of μ_2.

Hence the common multiples of a_1, a_2, a_3 ar the same as the multiples of μ_2.

Similarly the common multiples of a_1, a_2, a_3, a_4 ar the same as the multiples of μ_3. And so on.

Finally the common multiples of $a_1, a_2, a_3, \cdots, a_n$ ar the same as the multiples of μ_{n-1}, which is the least common multiple of $a_1, a_2, a_3, \cdots, a_n$.

719. Theorem. *If ν is a common multiple of the integers $a_1, a_2, a_3 \cdots, a_n$, then*

$$(a_1 \times a_2 \times a_3 \times \cdots \times a_n)$$
$$\times\ [(\nu : a_1) \otimes (\nu : a_2) \otimes (\nu : a_3) \otimes \cdots \otimes (\nu : a_n)]$$
$$= |\nu|,$$ *provided ν is not zero* (2).

By § 697 $(a_1 \times a_2)\ [(\nu : a_1) \otimes (\nu : a_2)] = |\nu|$.

That is, the theorem holds for two integers.

---

(1) See P. Barrieu, *Mathesis*, Vol. III., 1883, p. 217; Stieltjes, p. 2.

(2) See Barrieu, p. 217.

To prove it in general, suppose that it holds for k integers, wher $2 \gtreqless k < n$.

Then we hav

$$(a_1 \times a_2 \times a_3 \times \cdots \times a_k)$$
$$\times [(\nu : a_1) \otimes (\nu : a_2) \otimes (\nu : a_3) \otimes \cdots \otimes (\nu : a_k)]$$
$$= |\nu|.$$

Hence $(\nu : a_1) \otimes (\nu : a_2) \otimes (\nu : a_3) \otimes \cdots \otimes (\nu : a_k)$
$= |\nu| : (a_1 \times a_2 \times a_3 \times \cdots \times a_k)$.

Hence

$$(\nu : a_1) \otimes (\nu : a_2) \otimes (\nu : a_3) \otimes \cdots \otimes (\nu : a_k) \otimes (\nu : a_{k+1})$$
$$= (|\nu| : (a_1 \times a_2 \times a_3 \times \cdots \times a_k)) \otimes (\nu : a_{k+1})$$
$$= (\nu : (a_1 \times a_2 \times a_3 \times \cdots \times a_k)) \otimes (\nu : a_{k+1})$$
$$= |\nu| : ((a_1 \times a_2 \times a_3 \times \cdots \times a_k) \times a_{k+1}) \quad \S\ 697.$$
$$= |\nu| : (a_1 \times a_2 \times a_3 \times \cdots \times a_k \times a_{k+1}).$$

Therfor

$$(a_1 \times a_2 \times a_3 \times \cdots \times a_k \times a_{k+1})$$
$$\times [(\nu : a_1) \otimes (\nu : a_2) \otimes (\nu : a_3) \otimes \cdots$$
$$\otimes (\nu : a_k) \otimes (\nu : a_{k+1})]$$
$$= |\nu|.$$

If, then, the theorem holds for k integers, it holds also for $k + 1$.

Therfor by the law of induction it holds for n integers.

720. Theorem. *If* ν *is the least common multiple of the integers* $a_1, a_2, a_3, \ldots, a_n$, *then* $(\nu : a_1) \otimes (\nu : a_2) \otimes (\nu : a_3) \otimes \cdots \otimes (\nu : a_n) = 1$;

and conversely, if the positiv integer ν *is a common multiple of the integers* a_1, a_2, a_3, $\cdots$, a_n *and* $(\nu : a_1) \otimes (\nu : a_2) \otimes (\nu : a_3) \otimes \cdots \otimes (\nu : a_n) = 1$, *then* ν *is the least common multiple of* a_1, a_2, a_3, $\cdots$, a_n (1).

Both parts of this theorem follow easily from the preceding theorem.

721. This theorem is equivalent to the following, which is the analog of § 658.

Theorem. *Having given a series of integers* a_1, a_2, a_3, $\cdots$, a_n, *none of which is zero; if* μ *is their least common multiple, ther exists a series of integers* γ_1, γ_2, γ_3, $\cdots$, γ_n, *none of which is zero and whose greatest common factor is unity, such that* $\mu = a_1\gamma_1 = a_2\gamma_2 = a_3\gamma_3 = \cdots = a_n\gamma_n$; *and conversely, if* $\mu = a_1\gamma_1 = a_2\gamma_2 = a_3\gamma_3 = \cdots = a_n\gamma_n$, *wher none of the integers* γ_1, γ_2, γ_3, $\cdots$, γ_n *is zero and their greatest common factor is unity, then* μ *is the least common multiple of* a_1, a_2, a_3, $\cdots$, a_n.

722. Theorem. *Having given a series of n integers, none of which is zero, the product of their least common multiple and the greatest common factor of the n products, each having as factors all but one of the given integers, is equal to the numerical value of the product of the given integers* (2).

This follows from § 719 by supposing

$$\nu = a_1 a_2 a_3 \cdots a_n.$$

(1) See Barrieu, p. 217; Lucas, p. 345; Tannery, p. 124.

(2) See Lucas, p. 346; Barrieu, p. 217; Stieltjes, p. 2.

723. Theorem. *Having given a series of n integers, none of which is zero, the product of the least common multiple of all the products that can be formd, each having as factors k of these integers, wher $k < n$, and the greatest common factor of all the products that can be formd, each having as factors $n - k$ of these integers, is equal to the numerical value of the product of the given integers* (1).

Let $a_1, a_2, a_3, \cdots, a_n$ be the given integers and

$$\nu = a_1 a_2 a_3 \cdots a_n.$$

Let also $\beta_1, \beta_2, \beta_3, \cdots, \beta_m$ be the series of products of the a's taken k at a time.

Then ν is a common multiple of the β's and $\nu : \beta_1, \nu : \beta_2, \nu : \beta_3, \cdots, \nu : \beta_m$ is the series of products of the a's taken $n - k$ at a time.

Now

$$(\beta_1 \times \beta_2 \times \beta_3 \times \cdots \times \beta_m)$$

$$\times [(\nu : \beta_1) \otimes (\nu : \beta_2) \otimes (\nu : \beta_3) \otimes \cdots \otimes (\nu : \beta_m)] = |\nu|.$$

§ 719.

Hence the theorem is proved.

This theorem contains the preceding as a particular case.

724. Theorem. *If ν is a common multiple of the integers $a_1, a_2, a_3, \cdots, a_n$, then* $(a_1 \otimes a_2 \otimes a_3 \otimes \cdots \otimes a_n)$ $[(\nu : a_1) \times (\nu : a_2) \times (\nu : a_3) \times \cdots \times (\nu : a_n)] = |\nu|$, *provided ν is not zero.*

---

(1) See Lucas, p. 346; Barrieu, p. 217.

This theorem is proved from § 699 by the method by which § 719 was proved from § 697.

725. Theorem. *Having given a series of n integers, none of which is zero, the product of their greatest common factor and the least common multiple of the n products, each having as factors all but one of the given integers, is equal to the numerical value of the product of the given integers* (1).

This follows from § 724 by supposing

$$\nu = a_1 a_2 a_3 \cdots a_n.$$

Or this theorem may be proved from § 723 by supposing the k of that theorem equal to $n - 1$ and using the commutativ law for multiplication.

726. Theorem. *Having given a series of n integers, none of which is zero, the product of the greatest common factor of all the products that can be formd, each having as factors k of these integers, wher $k < n$, and the least common multiple of all the products that can be formd, each having as factors $n - k$ of these integers, is equal to the numerical value of the product of the given integers* (2).

This theorem may be proved in the same way as § 723.

Or it follows from § 723 by putting $n - k$ for k and using the commutativ law for multiplication.

---

(1) See Barrieu, p. 218 ; Stieltjes, p. 2.

(2) See Barrieu, p. 219.

727. Theorem. *If the integers a_1, a_2, a_3, $\cdots$, none of which is zero, ar prime to each other two by two, their least common multiple is $|a_1a_2a_3\cdots|$, and conversely* (1).

For the direct theorem, setting $\nu = |a_1a_2a_3\cdots|$, since a_1, a_2, a_3, $\cdots$ ar prime to each other two by two, $(\nu : a_1) \otimes (\nu : a_2) \otimes (\nu : a_3) \otimes \cdots = 1$. § 666.

Therfor ν is the least common multiple of a_1, a_2, a_3, $\cdots$ § 720.

For the converse theorem, since ν is the least common multiple of a_1, a_2, a_3, $\cdots$,

$$(\nu : a_1) \otimes (\nu : a_2) \otimes (\nu : a_3) \otimes \cdots = 1. \quad \S\ 720.$$

Therfor a_1, a_2, a_3, $\cdots$ ar prime to each other two by two. § 666.

728. Theorem. *If the integers a_1, a_2, a_3, $\cdots$, none of which is zero, ar prime to each other two by two, and θ is any integer not zero, the least common multiple of θa_1, θa_2, θa_3, $\cdots$ is $|\theta a_1a_2a_3\cdots|$, and conversely.*

This follows from the preceding theorem, since $(\theta a_1) \times (\theta a_2) \times (\theta a_3) \times \cdots = |\theta|(a_1 \times a_2 \times a_3 \times \cdots)$. § 713.

729. Theorem. *If μ is the positiv integer which has every prime factor that occurs in any one of a series of positiv integers, and which contains each prime factor a number of times equal to the greatest number of times that it occurs as a factor in the given integers, then μ is the least common multiple of those integers.*

First, since μ contains every prime factor that occurs in any one of the given integers a number of

(1) See Tannery, p. 121.

times not less than the number of times that it occurs in that integer, μ is a common multiple of the given integers. § 640.

Now let ν be any positiv common multiple of the given integers.

Then every prime factor of any one of the given integers must be a factor of ν and must occur in ν a number of times not less than the number of times that it occurs in any one of the given integers (§ 640), and hence not less than the greatest number of times that it occurs as a factor in the given integers.

Therfor every prime factor of μ occurs in ν a number of times not less than the number of times that it occurs in μ.

Hence ν is a multiple of μ.

Since, therfor, μ is a common multiple of the given integers and every positiv common multiple is a multiple of μ, μ is the least common multiple of the given integers.

CHAPTER X.

CONGRUENCE.

730. If μ is any given integer, β and φ any integers, and we set $\alpha = \beta + \varphi\mu$, then $\alpha - \beta = \varphi\mu$. By giving β and φ various values we may thus obtain any number of pairs of integers which differ by a multiple of μ. This is the justification for the existence of the following definition.

731. Definition. An integer α is **congruent** (1) with an integer β with respect to an integer μ, calld the **modulus of congruence**, when the difference $\alpha - \beta$ is a multiple of μ.

Thus, if $\alpha - \beta = \varphi\mu$, α is congruent with β with respect to the modulus μ.

The statement of congruence is written

$$\alpha \equiv \beta \text{ (mod. } \mu),$$

which is read "α is congruent with β, modulo μ."

The statement $\alpha \equiv \beta$ (mod. μ) is calld a **congruence**; α is its **left member** and β its **right.**

When ther is no doubt what the modulus is, or when several congruences ar supposed to hold with

(1) The idea of congruence is due to Gauss. See his "Disquisitiones Arithmeticæ," p. 1. See also Dirichlet-Dedekind, p. 33; Chrystal, "Text-Book of Algebra," Part II., p. 500; Tannery, p. 454.

respect to the same modulus, we may write simply $\alpha \equiv \beta$.

If α is not congruent with β with respect to a given modulus μ, we may write $\alpha \not\equiv \beta$ (mod. μ).

732. If $\alpha \equiv \beta$ (mod. 0), $\alpha - \beta = \varphi \cdot 0 = 0$ and $\alpha = \beta$; and conversely, if $\alpha = \beta$, $\alpha - \beta = 0 = \varphi \cdot 0$ and $\alpha \equiv \beta$ (mod. 0).

That is, if the modulus were zero, the congruence $\alpha \equiv \beta$ would be equivalent to the equality $\alpha = \beta$. The idea of congruence would therfor, with a zero modulus, be equivalent to the old idea of equality. It would be nothing new.

We will, therfor, in the future assume that the modulus is not zero.

733. Theorem. *Any integer is congruent with itself with respect to any modulus.*

For $\alpha - \alpha = 0 \cdot \mu$.

734. Theorem. *If* $\alpha = \beta$, *then* $\alpha \equiv \beta$ *with respect to any modulus.*

735. Theorem.

If $\alpha \equiv \beta$ *(mod.* μ*), then* $\beta \equiv \alpha$ *(mod.* μ*).*

For, if $\alpha - \beta = \varphi\mu$, $\beta - \alpha = (-\varphi)\mu$.

736. Theorem.

If $\alpha \not\equiv \beta$ *(mod.* μ*), then* $\beta \not\equiv \alpha$ *(mod.* μ*).*

737. Theorem. *Any two integers ar congruent with respect to unity as a modulus.*

For $\alpha - \beta = (\alpha - \beta) \times 1$.

738. Theorem.

If $\alpha \equiv \beta$ *(mod.* μ*), then* $\bar{\alpha} \equiv \bar{\beta}$ *(mod.* μ*).*

For, if $\alpha - \beta = \varphi\mu$, $\bar{\alpha} - \bar{\beta} = (-\varphi)\mu$.

739. Theorem.

If $\alpha \equiv \beta$ *(mod.* μ*), then* $\alpha \equiv \beta$ *(mod.* $\bar{\mu}$*).*

740. Theorem. *If* $\alpha \equiv \beta$ *(mod.* μ*) and* $\mu = \nu$*, then* $\alpha \equiv \beta$ *(mod.* ν*).*

For, if $\alpha - \beta = \varphi\mu$ and $\mu = \nu$, $\alpha - \beta = \varphi\nu$.

741. Theorem. *If* $\alpha \equiv \beta$ *(mod.* μ*) and*

$\beta \equiv \gamma$ *(mod.* μ*), then* $\alpha \equiv \gamma$ *(mod.* μ*).*

Since $\alpha \equiv \beta$, $\alpha - \beta = \varphi\mu$.

Since $\beta \equiv \gamma$, $\beta - \gamma = \chi\mu$.

Adding, $\alpha - \gamma = (\varphi + \chi)\mu$.

Therfor $\alpha \equiv \gamma$.

742. Theorem. *If* $\alpha \equiv \beta$ *and* $\beta \not\equiv \gamma$*, then* $\alpha \not\equiv \gamma$.

743. Theorem. *If we hav a series of congruences in which the right member of each congruence, except the last, is also the left member of the succeeding congruence, as* $\alpha \equiv \beta \equiv \gamma \equiv \delta \equiv \varepsilon \equiv \cdots$*, then any integer in the series is congruent with any other integer in the series.*

744. Theorem. *If* $\alpha \equiv \beta$*, then* $\alpha + \gamma \equiv \beta + \gamma$*, and conversely.*

For the direct theorem,

since $\alpha \equiv \beta$, $\alpha - \beta = \varphi\mu$.

Hence $(\alpha + \gamma) - (\beta + \gamma) = \varphi\mu$.

Therfor $\alpha + \gamma \equiv \beta + \gamma$.

For the converse theorem reverse the steps.

745. Theorem. *If* $\alpha \equiv \beta$, *then* $\gamma + \alpha \equiv \gamma + \beta$, *and conversely.*

746. Theorem. *If* $\alpha \equiv \beta$ *and* $\gamma \equiv \delta$, *then* $\alpha + \gamma \equiv \beta + \delta$; *and conversely, if* $\alpha + \gamma \equiv \beta + \delta$ *and* $\gamma \equiv \delta$, *then* $\alpha \equiv \beta$.

747. Theorem. *The sum of the left members of any number of congruences is congruent with the sum of the right members.*

748. Theorem. *If* $\alpha \equiv \beta$, *then* $\alpha - \gamma \equiv \beta - \gamma$, *and conversely.*

749. Theorem. *If* $\alpha \equiv \beta$, *then* $\gamma - \alpha \equiv \gamma - \beta$, *and conversely.*

750. Theorem. *If* $\alpha \equiv \beta$ *and* $\gamma \equiv \delta$, *then* $\alpha - \gamma \equiv \beta - \delta$; *and conversely, if* $\alpha - \gamma \equiv \beta - \delta$ *and* $\gamma \equiv \delta$, *then* $\alpha \equiv \beta$; *if* $\alpha - \gamma \equiv \beta - \delta$ *and* $\alpha \equiv \beta$, *then* $\gamma \equiv \delta$.

751. Theorem. *If* $\alpha \equiv \beta$, *then* $\alpha\gamma \equiv \beta\gamma$.

Since $\alpha \equiv \beta$, $\qquad \alpha - \beta = \varphi\mu.$

Hence $\qquad (\alpha - \beta)\gamma = (\varphi\mu)\gamma.$

Or $\qquad \alpha\gamma - \beta\gamma = (\varphi\gamma)\mu.$

Therfor $\qquad \alpha\gamma \equiv \beta\gamma.$

752. Theorem. *If* $\alpha \equiv \beta$, *then* $\gamma\alpha \equiv \gamma\beta$.

753. Theorem. *If* $\alpha \equiv \beta$ *and* $\gamma \equiv \delta$, *then* $\alpha\gamma \equiv \beta\delta$.

754. Theorem. *The product of the left members of any number of congruences is congruent with the product of the right members.*

755. Theorem. *If* $\alpha \equiv \beta$ *(mod.* μ*), then* $\alpha\gamma \equiv \beta\gamma$ *(mod.* $\mu\gamma$*), provided* $\gamma \neq 0$*; and conversely, if* $\alpha\gamma \equiv \beta\gamma$ *(mod.* $\mu\gamma$*), then* $\alpha \equiv \beta$ *(mod.* μ*).*

For the direct theorem,

since $\alpha \equiv \beta\,(\text{mod. } \mu)$, $\alpha - \beta = \varphi\mu$.

Hence $(\alpha - \beta)\gamma = (\varphi\mu)\gamma$.

Or $\alpha\gamma - \beta\gamma = \varphi(\mu\gamma)$.

Therfor $\alpha\gamma \equiv \beta\gamma\,(\text{mod. } \mu\gamma)$.

For the converse theorem reverse the steps.

756. Theorem. *If* $\alpha \equiv \beta$ *(mod.* μ*) and* $\mu \mathrel{\dot{>}} \gamma$*, then* $\alpha \equiv \beta$ *(mod.* $\mu : \gamma$*).*

Since $\alpha \equiv \beta\,(\text{mod. } \mu)$, $\alpha - \beta = \varphi\mu$.

Since $\mu \mathrel{\dot{>}} \gamma$, $\mu = \gamma\nu$ and $\mu : \gamma = \nu$.

Hence $\alpha - \beta = (\varphi\gamma)\nu$.

Therfor $\alpha \equiv \beta\,(\text{mod. } \nu)$.

Or $\alpha \equiv \beta\,(\text{mod. } \mu : \gamma)$.

This theorem may be stated: *If two integers ar congruent with respect to a given modulus, they ar also congruent with respect to any factor of that modulus as a modulus.*

757. Theorem. *If* $\alpha \equiv \beta$ *(mod.* μ*) and* α*,* β*, and* μ *ar all divisible by* γ*, then* $\alpha : \gamma \equiv \beta : \gamma$ *(mod.* $\mu : \gamma$*), and conversely.*

Since α, β, and μ ar all divisible by γ, $\alpha = \delta\gamma$, $\beta = \varepsilon\gamma$, and $\mu = \nu\gamma$.

Hence, since $\alpha \equiv \beta\,(\text{mod. } \mu)$,

$$\delta\gamma \equiv \varepsilon\gamma \text{ (mod. } \nu\gamma).$$

Therfor $\delta \equiv \varepsilon$ (mod. ν). § 755.

Or $\alpha : \gamma \equiv \beta : \gamma$ (mod. $\mu : \gamma$).

For the converse theorem reverse the steps.

758. Theorem. *If $\alpha \equiv \beta$ (mod. μ) and α and β ar both divisible by γ, then $\alpha : \gamma \equiv \beta : \gamma$ (mod. $\mu : (\mu \otimes \gamma)$).*

Since $\alpha \equiv \beta$ (mod. μ), $\alpha - \beta = \varphi\mu$.

Let $\mu \otimes \gamma = \kappa$.

Then $\mu = \nu\kappa$ and $\gamma = \delta\kappa$, wher $\nu \perp\!\!\!\perp \delta$. § 604.

Also $\alpha - \beta = \varphi\nu\kappa$.

Since $\alpha \gtrdot \gamma$ and $\beta \gtrdot \gamma$, $\alpha - \beta \gtrdot \gamma$ and

$$(\alpha - \beta) : \gamma = \alpha : \gamma - \beta : \gamma. \quad \text{§ 479.}$$

Since $\alpha - \beta \gtrdot \gamma$, $\varphi\nu\kappa \gtrdot \delta\kappa$

and $(\alpha - \beta) : \gamma = (\varphi\nu\kappa) : (\delta\kappa)$. § 469.

Hence $\varphi\nu \gtrdot \delta$ and $(\alpha - \beta) : \gamma = (\varphi\nu) : \delta$.

Since $\varphi\nu \gtrdot \delta$ and $\nu \perp\!\!\!\perp \delta$, $\varphi \gtrdot \delta$. § 623.

Hence $(\alpha - \beta) : \gamma = (\varphi : \delta)\nu$. § 467.

Or $\alpha : \gamma - \beta : \gamma = (\varphi : \delta)(\mu : \kappa)$.

Therfor $\alpha : \gamma \equiv \beta : \gamma$ (mod. $\mu : \kappa$).

Or $\alpha : \gamma \equiv \beta : \gamma$ (mod. $\mu : (\mu \otimes \gamma)$).

If $\mu \gtrdot \gamma$, $\mu \otimes \gamma = |\gamma|$.

Hence, in this case, $\alpha : \gamma \equiv \beta : \gamma$ (mod. $\mu : |\gamma|$).

And $\alpha : \gamma \equiv \beta : \gamma$ (mod. $\mu : \gamma$).

This theorem, therfor, when $\mu \gtrdot \gamma$, reduces to the direct part of § 757.

759. Theorem.

If $\alpha \equiv \beta$ *(mod.* μ*) and* α *and* β *ar both divisible by* γ, *wher* $\mu \amalg \gamma$, *then* $\alpha : \gamma \equiv \beta : \gamma$ *(mod.* μ*)* (1).

For, if $\mu \amalg \gamma$, $\mu \otimes \gamma = 1$.

760. Theorem. *Any integer is congruent with each of the remainders that it leaves when divided by the modulus.*

For, if $\alpha = \varphi\mu + \rho$, $\alpha - \rho = \varphi\mu$.

761. Theorem. *If* $\alpha \equiv \rho$ *(mod.* μ*), wher* $|\rho| < |\mu|$, *then* ρ *is one of the remainders obtaind by dividing* α *by* μ. § 521.

762. Theorem. *If two integers ar congruent, they leave the same remainder when divided by the modulus, and conversely.*

For, if β leaves the remainder ρ, $\beta \equiv \rho$.

Hence, if $\alpha \equiv \beta$, $\alpha \equiv \rho$, wher $|\rho| < |\mu|$.

Therfor, when α is divided by μ, a remainder is ρ. § 761.

Conversely, if α and β both leave the remainder ρ, $\alpha \equiv \rho$ and $\beta \equiv \rho$.

Therfor $\alpha \equiv \beta$.

763. Theorem. *If two integers ar congruent and either is divisible by the modulus, the other is also; and conversely, if each of two integers is divisible by the modulus, they ar congruent.*

764. Theorem. *If two integers ar congruent and either is divisible by a factor of the modulus, the other is also.*

---

(1) See Tannery, p. 466.

765. Theorem. *If μ is any given modulus, all the integers, when written in the order of increasing magnitude, may be divided into successiv groups of* $|\mu|$ *each, so that the integers in each of these groups, taken in order, ar congruent with* $0, 1, 2, 3, \cdots, |\mu| - 1$.

The first integers in the successiv groups will be the various multiples of μ

$$\cdots, -3\mu, -2\mu, -\mu, 0, \mu, 2\mu, 3\mu, \cdots,$$
if μ is positiv,

$$\cdots, 3\mu, 2\mu, \mu, 0, -\mu, -2\mu, -3\mu, \cdots,$$
if μ is negativ.

If $\varphi\mu$ is either of these multiples, the group of which it is the first integer will be

$$\varphi\mu, \varphi\mu + 1, \varphi\mu + 2, \varphi\mu + 3, \cdots, \varphi\mu + (|\mu| - 1).$$

These integers ar in each case evidently congruent with $0, 1, 2, 3, \cdots, |\mu| - 1$.

766. Definition. This principle is calld the **periodicity** of the integers with respect to a given modulus.

767. Definition. A **cyclical arrangement** of any set of things, given in a certain order, as, for example, the letters *a, b, c, d, e, f*, is an arrangement obtaind by dividing the set into two parts at any place and then making the first part the second and vice versa, the order of the symbols in each part being unchanged.

Thus *b, c, d, e, f, a* and *e, f, a, b, c, d* ar cyclical arrangements of the above set of letters.

The original set unchanged is also said to be a cyclical arrangement of itself.

768. Theorem. *If* μ *is any given modulus and any group of* $|\mu|$ *successiv integers ar taken, they ar congruent, in order, with some cyclical arrangement of the integers* $0, 1, 2, 3, \cdots, |\mu|-2, |\mu|-1$.

Let $\varphi\mu+\rho$ be the first of the given group of integers, wher ρ is positiv and numerically less than μ. § 522.

Then the group of integers will be

$$\varphi\mu+\rho,\ \varphi\mu+\rho+1,\ \varphi\mu+\rho+2,\ \cdots,\\ \varphi\mu+\rho+(|\mu|-2),\ \varphi\mu+\rho+(|\mu|-1).$$

These differ each by a multiple of μ from the corresponding integers of the group

$$\rho,\ \rho+1,\ \rho+2,\ \cdots,\ \rho+(|\mu|-2),\ \rho+(|\mu|-1).$$

Therfor the integers of the first group ar respectivly congruent with these integers.

Now set $|\mu|=\rho+\kappa$, wher κ is positiv.

Then the latter group may be written

$$\rho,\ \rho+1,\ \rho+2,\ \cdots,\ \rho+(\kappa-2),\ \rho+(\kappa-1),\ \rho+\kappa,\\ \rho+(\kappa+1),\ \cdots,\ \rho+(|\mu|-2),\ \rho+(|\mu|-1),$$

or

$$\rho,\ \rho+1,\ \rho+2,\ \cdots,\ |\mu|-2,\ |\mu|-1,\ |\mu|,\ |\mu|+1,\\ \cdots,\ |\mu|+(\rho-2),\ |\mu|+(\rho-1).$$

These ar congruent respectivly with

$$\rho,\ \rho+1,\ \rho+2,\ \cdots,\ |\mu|-2,\ |\mu|-1,\ 0,\ 1,\ \cdots,\ \rho-2,\ \rho-1.$$

Therfor the original group of integers ar congruent respectivly with these integers, which ar a cyclical arrangement of the integers

$$0, 1, 2, 3, \cdots, |\mu| - 2, |\mu| - 1.$$

769. Theorem. *Any group of* $|\mu|$ *successiv integers ar congruent, in order, with respect to the modulus* μ, *with some cyclical arrangement of the set of* $|\mu|$ *integers obtaind from the set of integers* $0, 1, 2, 3, \cdots, |\mu| - 3, |\mu| - 2, |\mu| - 1$ *by replacing any or all of them by the corresponding integers of the set* $0, -(|\mu| - 1), -(|\mu| - 2), -(|\mu| - 3), \cdots, -3, -2, -1.$

For, if ρ is the positiv remainder obtaind by approximately dividing a given integer by μ, and σ the corresponding negativ remainder, $\sigma = -(|\mu| - \rho)$.

§ 518.

770. Theorem. In particular, *any group of* $|\mu|$ *successiv integers ar congruent, in order, with the integers*

$$-a, -(a-1), -(a-2), \cdots, -3, -2, -1, 0, 1, 2, 3, \cdots, |\mu| - a - 2, |\mu| - a - 1,$$

wher a *is either zero or positiv and less than* $|\mu|$.

FUNDAMENTAL IDEAS.

1. Number. § 1.
2. Equality. § 5.
3. Sum. § 37.

AXIOMS.

1. $a = a$. § 6.
2. If $a = b$, then $b = a$. § 7.
3. If $a = b$ and $b = c$, then $a = c$. § 34.
4. $a + b = b + a$. § 41.
5. If $a = b$, then $a \neq c + b$. § 43.
6. If $a \neq b$, ther is some number c, such that either $a = c + b$ or $c + a = b$. § 45.
7. If $a = b$, then $a + c = b + c$. § 46.
8. If $a \neq b$, then $a + c \neq b + c$. § 47.
9. $(a + b) + c = a + (b + c)$. § 54.
10. Every number, except one, is the sum of two numbers. § 61.

The author does not include § 40 in this list of axioms, as he has become convinced, since the electrotyping of most of the book that § 40 is not properly an axiom. That the number of objects in the group obtaind by joining two groups should be unique is a prerequisit to the definition of the symbol $a + b$. If this number were not one-valued, $a + b$ would not be defined. See § 416.

LIST OF AUTHORS AND BOOKS REFERD TO IN THE PRECEDING PAGES.

ARCHIMEDES.
W. W. R. BALL, A Short Account of the History of Mathematics, 1893.
P. BARRIEU, Mathesis, 1883.
BEMAN AND SMITH, Elements of Algebra, 1900; New Plane and Solid Geometry, 1901.
BHASKARA.
G. CANTOR, Sur les Fondements de la Théorie des Ensembles Transfinis, tr. from the German by F. Marotte, 1899.
A. CAYLEY.
G. CHRYSTAL, Text-Book of Algebra, 1893.
A. C. CLAIRAUT.
L. CREMONA, Elements of Projective Geometry, tr. from the Italian by C. Leudesdorf, 1885.
E. W. DAVIS, Logic of Algebra, 1890.
R. DESCARTES, La Géométrie, 1637.
U. DINI, Grundlagen für eine Theorie der Functionen einer veränderlichen reelen Grösse, tr. from the Italian by Lüroth and Schepp, 1892.
DIRICHLET-DEDEKIND, Vorlesungen über Zahlentheorie, 4th ed., 1894.
ERATOSTHENES.
EUCLID, Elements of Geometry.
L. EULER, Elements d'Algèbre, ed. of 1807.
FISHER AND SCHWATT, Text-Book of Algebra, Part I., 1898.
K. F. GAUSS, Disquisitiones Arithmeticæ, 1801.
J. WILLARD GIBBS, Vector Analysis, 1884.
A. GIRARD, Invention nouvelle en l'algèbre, 1629.
D. F. GREGORY, On a Difficulty in the Theory of Algebra, 1840.
J. HADAMARD, Leçons de Geométrie élémentaire, 1898.
WM. ROWAN HAMILTON, Transactions of the Royal Irish Academy, Vol. XVII., 1835; Taylor's Philosophical Magazine, 1844; Lectures on Quaternions, 1853.

H. HANKEL, Theorie der komplexen Zahlsysteme, 1867.
T. HARRIOT, Artis Analyticæ Praxis, 1631.
D. HILBERT, The Foundations of Geometry, tr. from the German by E. J. Townsend, 1902; Bulletin of the American Mathematical Society, Vol. VIII., 1902.
E. V. HUNTINGTON, Transactions of the American Mathematical Society, Vol. III., 1902.
JEVONS-HILL, Elements of Logic, 1883.
V. A. LEBESGUE, Théorie des Nombres, 1862.
A. LEFEVRE, Number and its Algebra, 1896.
A. M. LEGENDRE, Théorie des Nombres, 4th ed., 1900.
LEONARDO of Pisa, Liber Abaci, 1202.
J. LOCKE, An Essay Concerning Human Understanding, 1690.
E. LUCAS, Théorie des Nombres, 1891.
E. H. MOORE, Transactions of the American Mathematical Society, 1902.
JORDANUS NEMORARIUS, Algorithmus Demonstratus, 1534.
W. OUGHTRED, Clavis Mathematica Denuo Limata, 1648.
G. PEACOCK, Symbolical Algebra, 1845.
R. M. PIERCE, Problems of Number and Measure, 1898.
R. RECORD, Whetstone of Witte, 1557.
E. SCHRÖDER, Abriss der Arithmetik und Algebra, 1874.
H. SCHUBERT, Arithmetik und Algebra, 1899; Encyklopädie der Mathematischen Wissenschaften, Vol. I., 1898.
J. A. SERRET, Traité d'Arithmétique, 1852.
J. F. SERVOIS, Gergonne's Annales, Vol. V., 1814.
T. J. STIELTJES, Annales de la Faculté des Sciences de Toulouse, Vol. IV., 1890.
M. STIFEL, Arithmetica Integra, 1544.
STOLZ UND GMEINER, Theoretische Arithmetik, 1900.
W. SWINTON, New English Grammar, 1881.
J. TANNERY, Leçons d'Arithmétique, 1900.
F. VIETA, In Artem Analyticam Isagoge, 1591.
K. WEIERSTRASS, Gesammelte Werke, Vol. I.
G. A. WENTWORTH, Plane and Solid Geometry, 1901.
J. WIDMAN, Mercantile Arithmetic, 1489.

INDEX.

[The references ar to sections.]

Printed in Dunstable, United Kingdom

91219914R00157